Springer Series in Chemical Physics

Volume 117

Series editors

Albert W. Castleman, University Park, USA
Jan Peter Toennies, Göttingen, Germany
Kaoru Yamanouchi, Tokyo, Japan
Wolfgang Zinth, München, Germany

The purpose of this series is to provide comprehensive up-to-date monographs in both well established disciplines and emerging research areas within the broad fields of chemical physics and physical chemistry. The books deal with both fundamental science and applications, and may have either a theoretical or an experimental emphasis. They are aimed primarily at researchers and graduate students in chemical physics and related fields.

More information about this series at http://www.springer.com/series/676

Katharina Brinkert

Energy Conversion in Natural and Artificial Photosynthesis

 Springer

Katharina Brinkert
Division of Chemistry
  and Chemical Engineering
California Institute of Technology
Pasadena, CA
USA

and

European Space Agency
ESTEC
Noordwijk
The Netherlands

ISSN 0172-6218
Springer Series in Chemical Physics
ISBN 978-3-319-77979-9   ISBN 978-3-319-77980-5   (eBook)
https://doi.org/10.1007/978-3-319-77980-5

Library of Congress Control Number: 2018934942

Printed on acid-free paper

This Springer imprint is published by the registered company Springer International Publishing AG
part of Springer Nature
The registered company address is: Gewerbestrasse 11, 6330 Cham, Switzerland

Contents

1 Introduction 1

2 Oxygenic Photosynthesis—A Brief Overview 3
  2.1 The Light Reactions 3
  2.2 The Dark Reactions 6
  References 7

**3 Architecture, Structure and Function of the Energy
  Conversion Centers** 9
  3.1 Overview of the Evolution and Structure of the Photosynthetic
      Reaction Centers 9
      3.1.1 Antenna System Organization 10
      3.1.2 Reaction Center Structures and Cofactor
            Arrangement 14
  3.2 Semiconductor Photochemistry and Principles
      of Photoelectrocatalysis: Energetics, Charge Separation
      and Excitation Transfer 20
      3.2.1 Photoelectrochemistry of Semiconductors 23
      3.2.2 Photoelectrode Materials 25
  References 29

4 Electron Transfer 33
  4.1 Electron Transfer, Bioenergetics and Redox Potentials
      of Cofactors in Photosystem II 33
  4.2 Proton-Coupled Electron Transfer in Photosystem II:
      Kinetics and Thermodynamics 36
      4.2.1 The Tyrosyl Radicals in Photosystem II:
            Why D and Why Z? 39
  4.3 Back-Reactions, Short-Circuits and Leaks: Kinetics
      and Pathways of Charge Recombination 41

4.4   Photoelectrochemistry in Semiconductors: Electron
      Transfer and Recombination Reactions.   43
      4.4.1   Nanostructured Photoelectrodes   50
References   52

5   **Water Oxidation Catalysis and Hydrogen Evolution**   55
    5.1   Structure and Function of Nature's Oxygen Evolving
          Complex   56
          5.1.1   Channel Architecture   56
          5.1.2   Catalytic Cycle and Manganese Oxidation States   58
    5.2   Photoelectrocatalysis for Solar Water-Splitting   62
          5.2.1   Electrocatalysts: Oxygen Evolution on Semiconductor
                  Photoelectrodes   63
    5.3   Molecular Catalysts for Water Oxidation   65
    5.4   Photoelectrochemical Hydrogen Production   69
          5.4.1   Catalyst Materials and Reaction Mechanisms.   70
    References   72

6   **Carbon Fixation**   75
    6.1   RubisCO—Structure, Functionality and Catalytic
          Efficiencies.   76
    6.2   Photoelectrochemical CO_2 Reduction: Materials,
          Challenges and Strategies   79
          6.2.1   Sequential Versus Concerted Proton-Coupled
                  Electron Transfer Mechanisms   82
    References   84

7   **Protection Mechanisms**   87
    7.1   Photodamage in Photosystem II.   88
          7.1.1   Protection Against Light Damage   89
    7.2   Catalyst Stability and Photocorrosion in Photoelectrochemical
          Cells   92
          7.2.1   Role and Function of Protection Layers   92
    References   95

8   **Biomimetic Systems for Artificial Photosynthesis**   97
    8.1   Photosynthetic Model Systems: Concepts and Ideas
          to Realize Artificial Photosynthesis   97
          8.1.1   Electron Transfer Studies   99
          8.1.2   Proton-Coupled Electron Transfer Studies   102
          8.1.3   Synthetic Mimics of the Mn_4CaO_5 Cluster   103
    8.2   The Photoelectrochemical Device: Utilizing Solar Energy
          for Water Oxidation, Hydrogen Production and CO_2 Fixation . . .   105
    References   108

9   Efficiency of Photosynthesis and Photoelectrochemical Cells 111
     9.1   The Photosynthetic Efficiency 111
     9.2   Optimizing Photosynthesis? 113
     9.3   Efficiencies of Solar Water Splitting Cells 115
           9.3.1   Calculation of Solar-to-Chemical Conversion
                   Efficiencies 117
     9.4   Comparison of Photosynthetic and Photoelectrochemical
           Cell Efficiency? 119
     References 121

10   Conclusion: Towards the Realization of an Artificial Leaf 123

Index 125

Chapter 1
Introduction

About 2.3 billion years ago, nature developed its most important energy conversion process: photosynthesis. Since decades, the crucial elements and principles of converting water and carbon dioxide into oxygen and chemical energy using sunlight have inspired scientists in the design of artificial biomimetic systems: collecting and storing solar energy in chemical bonds is a highly desirable approach to solving the energy challenge. Artificial systems for solar-driven water-splitting face, however, similar challenges to biological photosynthesis: efficient mechanisms for light absorption, charge separation and transfer, catalysis and protection against corrosion are required to provide a stable and constant energy flux from the photoelectrolysis of water.

This monograph aims at discussing the basic principles and processes of natural photosynthesis and comparing them directly with recent developments and concepts currently realized in artificial photosynthetic systems, capable of utilizing sunlight and converting carbon dioxide and water into a chemical fuel. Here, the main focus lies on photoelectrochemical cells, where semiconducting photoanodes and -cathodes modified with (electro-) catalysts are used to oxidize water, produce hydrogen and reduce carbon dioxide in a monolithic device. Fundamental photochemical and photophysical processes are presented and discussed along with protection mechanisms and efficiency calculations for both, natural and artificial photosynthesis. Key parameters are identified which are crucial for the efficient operation of natural photosynthesis and their validity and applicability for the design of artificial solar-driven water-splitting systems are further on examined.

© Springer International Publishing AG, part of Springer Nature 2018
K. Brinkert, *Energy Conversion in Natural and Artificial Photosynthesis*,
Springer Series in Chemical Physics 117,
https://doi.org/10.1007/978-3-319-77980-5_1

Chapter 2
Oxygenic Photosynthesis—A Brief Overview

Oxygenic photosynthesis is the process in which oxygen and carbohydrates are produced from water and carbon dioxide utilizing sunlight. With the ability of converting solar energy into chemical energy, oxygenic photosynthesis remains one of the most important biological processes on earth, producing directly or indirectly the building blocks of all living organisms and also of oil, gas and coal. In contrast to some anaerobic photosynthetic organisms which use H_2S as a reducing agent and produce sulphur as a major product, oxygenic photosynthesis utilizes H_2O as the reducing agent. It is accomplished by a series of reactions which occur in the chloroplast of photosynthetic eukaryotes and cyanobacteria and is characterized by two main sets of reactions: the 'light' and the 'dark' reactions. Early biochemical studies demonstrated that chloroplast thylakoid membranes oxidize H_2O, reduce NADP to NADPH and synthesize ATP during the so-called 'light reactions' [1]. These reactions were shown to be catalyzed by two photosystems, Photosystem I (PSI) and Photosystem II (PSII), an ATP synthase, which produces the energy source ATP through a proton gradient formed by light-driven electron transfer reactions and the cytochrome(cyt) b_6f complex, mediating the electron transport between PSII and PSI and partly converting the redox energy into the proton gradient used for ATP formation [2]. The 'dark' reactions utilize the reducing equivalents NAPDH and ATP generated by the light reactions to reduce atmospheric CO_2 into carbohydrates, which act as the respiratory source.

2.1 The Light Reactions

For the light-driven reactions in photosynthesis, four protein complexes are required which reside in the thylakoids, a membrane continuum of flattened sacs in the inner membrane part of the chloroplast [3]. They form a three-dimensional network enclosing an aqueous space called the lumen and are differentiated into cylindrical stacked structures, the grana, and interconnecting single membrane

© Springer International Publishing AG, part of Springer Nature 2018
K. Brinkert, *Energy Conversion in Natural and Artificial Photosynthesis*,
Springer Series in Chemical Physics 117,
https://doi.org/10.1007/978-3-319-77980-5_2

regions, the stromal lamellae (Fig. 2.1). The electron transfer and energy trans-duction are catalyzed by proteins which are distributed in the thylakoid membrane: the F-ATPase is located mainly in the stromal lamellae, the cyt b_6f complex can be found in the grana and grana margins, PSI is located in the stromal lamellae and PSII is found in the grana, [4, 5]. The outer membrane of the chloroplast encloses the stromal compartment which contains the soluble enzymes required for the dark reactions and in addition to enzymes that synthesize proteins and the genome.

The accepted model of photosynthetic electron transport during the light reactions is illustrated in Fig. 2.2, following a so-called 'Z-scheme' (Fig. 2.3): Photosystem II absorbs light energy through an assembly of light-harvesting chlorophylls, funneling the trapped light energy to the central chlorophylls in the reaction center. This induces the formation of an excited electronic state over the central pigments, which are known as P_{680} according to the wavelength of their lowest-energy absorption band. P_{680}^* is a strong reducing agent and rapidly loses an electron to a nearby electron acceptor (A). A cascade of fast electron transfer reactions stabilizes this charge separation by increasing the distance between the electron donor and acceptor and therefore, reducing recombination reactions. The electron is passed along an electron transport chain via the cyt b_6f complex to Photosystem I (PSI). PSI simultaneously absorbs light energy through light-harvesting chlorophylls and generates an electron for the reduction of NADP

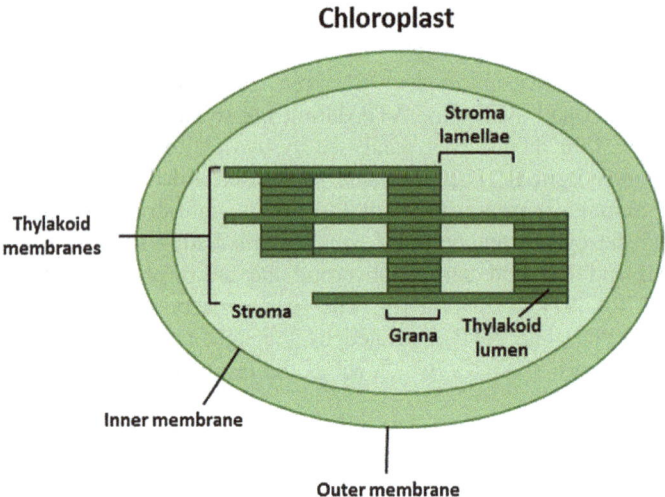

Fig. 2.1 Chloroplasts are organelles which are found in the cells of green plants and cyanobacteria. They are bound by a double membrane and contain a third membrane system which are known as thylakoids. The thylakoid membranes of higher plants consist of stacks of membranes which form the granal-regions, connected to adjacent grana by non-stacked membranes called stroma-lamellae. The compartment surrounding the thylakoids is known as stroma, the space inside the thylakoids as lumen

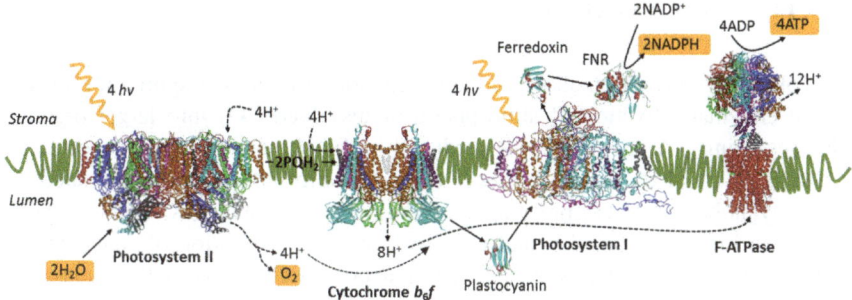

Fig. 2.2 Schematic overview of the electron transport chain of the photosynthetic light reaction

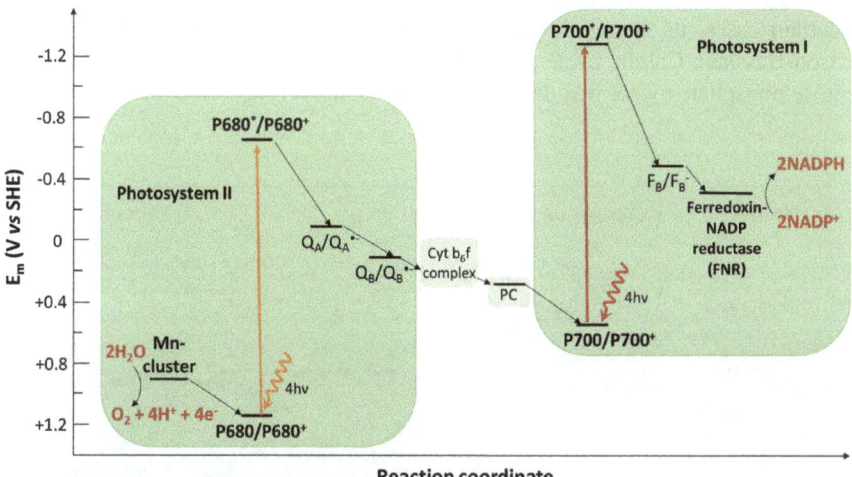

Fig. 2.3 Energetic scheme of the photosynthetic light reaction ('Z-scheme'). Indicated are the redox potentials of the most important cofactors

to NADPH by the ferredoxin-NADP reductase (FNR). PSI is reduced again by the electron originating from the light absorption process in PSII. In turn, the remaining cation in PSII is able to oxidize a tyrosine residue and subsequently, a pentanuclear tetramanganese-calcium cluster (Mn_4CaO_5), commonly referred to as the oxygen-evolving complex (OEC). The catalytic cycle of the OEC involves five different oxidation states (S_i, $i = 0$–4), generating electrons, protons and oxygen from the oxidation of water [6]:

$$2H_2O \rightarrow 4H^+ + 4e^- + O_2 \qquad (2.1)$$

2.2 The Dark Reactions

The evolutionary invention allowing the production of biomass required for all life was the enzymatic fixation of atmospheric or dissolved CO_2 into larger organic molecules using NADPH and ATP produced in the photosynthetic light reaction. According to fossil records of bacterial biofilms, unicellular organisms related to modern cyanobacteria were presumably among the first organisms which coupled light driven biomass formation and CO_2 fixation to the oxidation of water which resulted in the release of oxygen [7]. The reaction mechanisms of the photosynthetic carbon assimilation (also light-independent reactions or 'dark reactions') were firstly recognized by three American scientists, Andrew A. Benson, James A. Bassham and Melvin Calvin who unveiled the biochemical reaction sequence which fix and reduce atmospheric carbon into organic matter and therefore, satisfies the energy needs and demands of phototrophic cells, heterotrophic plant tissues or in certain cases, its symbiotic partner [8]. In a short period of time [9–13], the Benson-Bassham-Calvin cycle (named in honor of its discovers) or reductive pentose phosphate cycle, was developed by the authors based on experiments with

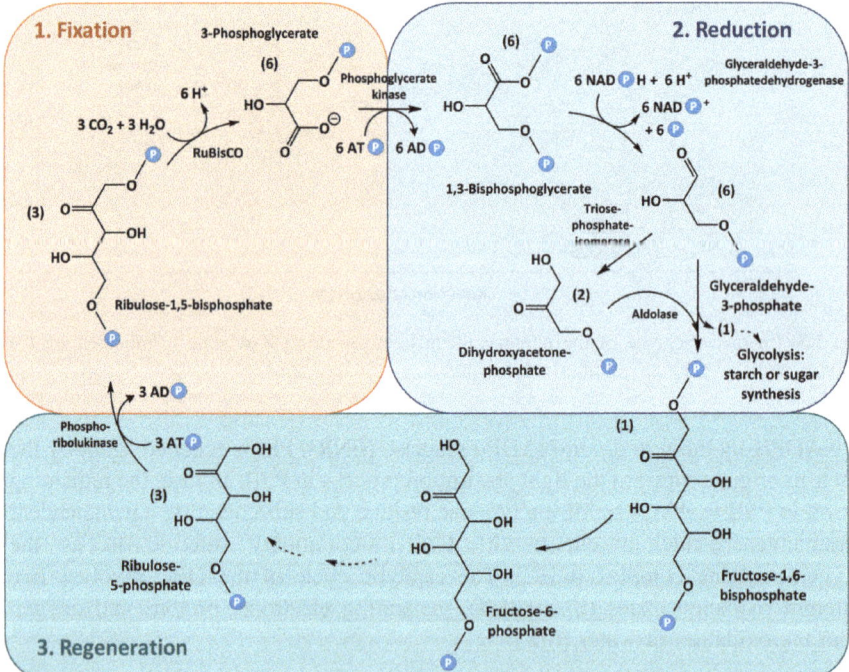

Fig. 2.4 The three stages of CO_2 assimilation in photosynthetic organisms in the carbon reduction cycle (Benson-Bassham-Calvin cycle). As shown here, for the net synthesis of one molecule of glyceraldehyde-3-phosphate, three CO_2 molecules are fixed. It is used further on for the production of starch or sugars. The ribulose-5-phosphate generation is shown exemplarily via the fructose-1,6-bisphosphate pathway

microalgae. Radioactive substrates and chromatographic analysis were used to study biochemical pathways. They could prove that primary CO_2 photoassimilation occurs in the carboxylation reaction of ribulose-1,5-bisphosphate (RubP) by accepting a molecule CO_2 (Fig. 2.4). The first stable product is 3-phosphoglyceric acid (3-PGA) which is then further on reduced to glyceraldehyde-3-phosphate and dihydroxyacetone phosphate (DOPA). The triose phosphates yield hexose phosphate esters, used for the consecutive synthesis of sucrose and/or starch. The interconversion of phosphoric sugar esters in the sequence of transketolase and aldolase reactions regenerates RubP. Furthermore, Ribulose-5-phosphate is phosphorylated again by the phosphoribulokinase using 3 ATP.

Every year, photosynthetic organisms convert in total about 10^{17} g (100 Gt) of CO_2 into biomaterials and organic compounds, whereas all oxygenic photosynthetic organisms utilize the enzyme RuBisCO (ribulose-1,5-bisphosphat-carboxylase/-oxygenase) for the initial carbon fixation reaction (see Chap. 6) [14]. Although, different terrestrial environments caused structural and functional adaptations throughout the plant body, the Benson-Bassham-Calvin cycle has not changed its fundamental biochemical nature over the cause of evolution and is well protected throughout photosynthetic organisms.

References

1. L.P. Vernon, M. Avron, Photosynthesis. Annu. Rev. Biochem. **34**, 269–296 (1965)
2. N. Nelson, C.F. Yocum, Structure and function of photosystems I and II. Annu. Rev. Plant Biol. **57**, 521–565 (2006)
3. N. Nelson, A. Ben-Shem, The complex architecture of oxygenic photosynthesis. Nat. Rev. Mol. Cell Bio. **5**, 971–982 (2004)
4. P.A. Albertsson, A quantitative model of the domain structure of the photosynthetic membrane. Trends Plant Sci. **6**, 349–358 (2001)
5. J.M. Anderson, Changing concepts about the distribution of photosystems I and II between grana-appressed and stroma exposed thylakoid membranes. Photosynth. Res. **73**, 157–164 (2002)
6. B. Kok, B. Forbush, M. McGloin, Cooperation of charges in photosynthetic O_2 evolution–I. A linear four step mechanism. Photochem. Photobiol. **11**, 467–475 (1970)
7. B. Rasmussen, I.R. Fletcher, J.J. Brocks, M.R. Kilburn, Reassessing the first appearance of eukaryotes and cyanobacteria. Nature **455**, 1101–1104 (2008)
8. K. Biel, I. Fomina, Benson-Bassham-Calvin cycle contribution to the organic life on our planet. Photosynthetica **53**(2), 161–167 (2015)
9. M. Calvin, A.A. Benson, The path of carbon in photosynthesis. Science **107**, 476–480 (1948)
10. A.A. Benson, M. Calvin, Carbon dioxide fixation by green plants. Annu. Rev. Plan Phys. **1**, 25–42 (1950a)
11. A. A. Benson, M. Calvin, The path of carbon in photosynthesis: VII. Respiration and photosynthesis. J. Ex. Bot. **1**, 63–68 (1950b)

12. A.A. Benson, J.A. Bassham, M. Calvin, T.C. Goodale, V.A. Haas, W. Stepka, The path of carbon in photosynthesis. V. Paper chromatography and radioautography of the products. J. Am. Chem. Soc. **72**(4), 1710–1718 (1950)
13. J.A. Bassham, A.A. Benson, L.D. Kay, A.Z. Harris, A.T. Wilson, M. Calvin, The path of carbon in photosynthesis. XXI. The cyclic regeneration of carbon dioxide acceptor. J. Am. Chem. Soc. **76**(7), 1760–1770 (1954)
14. P.G. Kroth, The biodiversity of carbon assimilation. J. Plant Physiol. **172**, 76–81 (2015)

Chapter 3
Architecture, Structure and Function of the Energy Conversion Centers

The sun provides more than 100,000 TW to the surface of the earth. The evolution of photosynthesis occurred based on maximizing the productive process of generating new components and energy for biological activities. Today, photosynthesis provides more than 150 TW of chemical energy on land and in the oceans, which is about 10 times more than humans' global consumption of primary energy [1]. In most of the photosynthetic processes, photosynthetic reaction centers convert light energy into chemical energy. They operate in organisms ranging from bacteria to higher plants and are evolutionary linked, which is demonstrated by partial amino acid sequence homologies and their fine structure [2]. The core of every photochemical reaction center is a dimeric structure. It is assumed that its evolution began with a homodimeric structure and progressed from symmetric via pseudosymmetric to asymmetric structures [3].

There are two main types of reaction centers which are differentiated by their type of electron acceptors: in type I reaction centers, excited electrons are captured by electron acceptors such as ferredoxin; in type II reaction centers, excited electrons are captured by a loosely bound quinone. The type I reaction centers include the Photosystem Is (PSI) of oxygenic photosynthesis and bacteria phyla, such as the green sulfur bacteria. Type II reaction centers include the Photosystem IIs (PSII) of oxygenic photosynthesis and bacteria such as purple bacteria.

3.1 Overview of the Evolution and Structure of the Photosynthetic Reaction Centers

The determination of the reaction centers' origin poses a challenge due to their diversity, but the greater availability of genomic sequences and of high-resolution crystal structures from various organisms enable the determination of their common evolutionary path. Due to the large energy difference in the redox potential between

© Springer International Publishing AG, part of Springer Nature 2018 9
K. Brinkert, *Energy Conversion in Natural and Artificial Photosynthesis*,
Springer Series in Chemical Physics 117,
https://doi.org/10.1007/978-3-319-77980-5_3

the electron donor (water) and the final electron acceptor ($NADP^+$) during the light reaction, the ancestor cyanobacteria had to evolve the capability to use two photosystems working in series in order to be able to accumulate the energy of two photons. About 1.5 billion years ago, the ancestral symbiotic event between a cyanobacterium and a eukaryotic cell transformed the first organism into a proto-chloroplast and opened the way to the evolution of green eukaryotic photosynthetic organisms i.e., plants and green algae [4]. For almost 2 billion years, life remained mainly in water; land plants appeared only about 0.5 billion years ago. The first land plants had to challenge new environmental constraints such as an oxygen rich atmosphere, a rapidly fluctuating environment in terms of light quantity and quality, temperature, nutrients and water [5].

The two photosystems have a common organization and are functionally organized in two main moieties: a core complex, containing the reaction center for the photochemical processes and a peripheral antenna system, increasing the light harvesting capability and regulating the photosynthetic process [6]. Most of the subunits are similar in prokaryotic and eukaryotic photosystems, indicating that the two core complexes have been well conserved during evolution in all oxygen-evolving organisms [7]. The peripheral antenna system, on the contrary, shows a great variability and composes peripheral associated membrane proteins in cyanobacteria, phycobilisomes and integral membrane proteins which compose the light harvesting complexes (LHCs) in eukaryotic cells.

3.1.1 Antenna System Organization

Charge separation takes place in pigment-protein complexes referred to as the reaction centers (RCs). They are highly specialized and have a low pigment density which leads to little light absorption. Light-harvesting complexes surround the RCs, also called 'antennas', which contain a few hundred pigments per RC (Fig. 3.1a–c) [8]. These complexes are crucial for the success of photosynthesis due to the fact that light is dilute. Even on a sunny day, a chlorophyll molecule will absorb not more than one photon every 0.1 s [9]. Energy transfer to the reaction center (RC) core pigments of PSII reaches remarkable efficiencies of up to 90% [10]. The reaction centers would be inactive most of the time without the antennae, leading to the loss of excitation energy. Therefore, the ability of harvesting light is crucial for the organism in order to catch every available photon in light-limited conditions. Furthermore, as light quality and quantity changes, the antenna system represents a unit, which can be designed ad hoc: the LHCs show a remarkable variability in pigment composition, pigment organization and size of the antenna system in different natural environments. They can also prevent the overexcitation of the photosynthetic machinery, including photodamage. How do they manage?

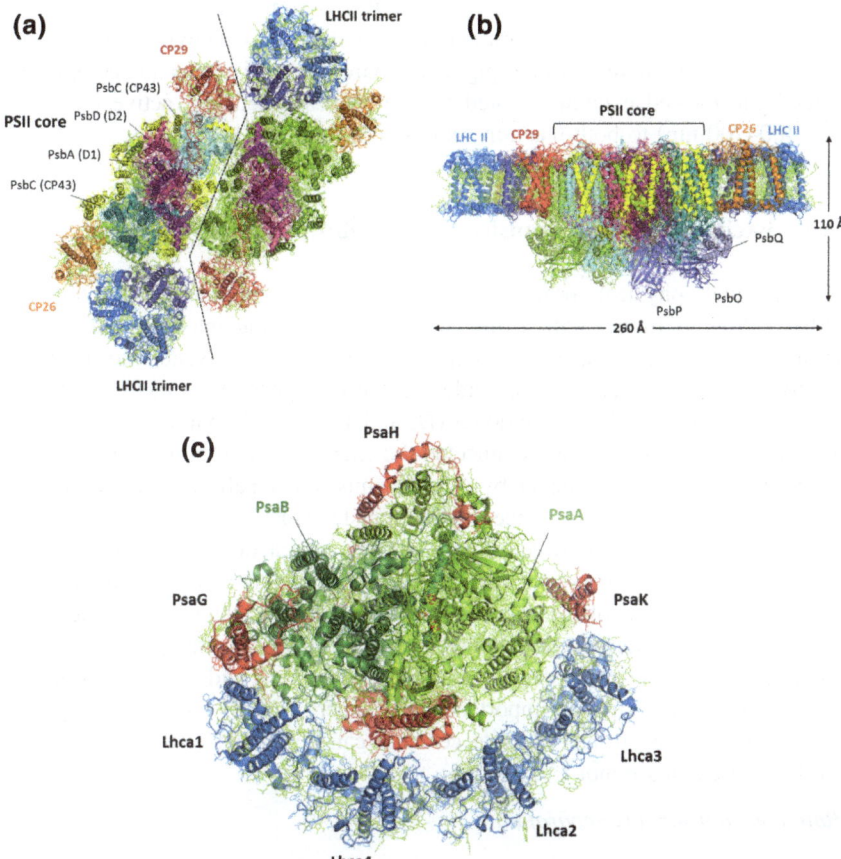

Fig. 3.1 **a** and **b** Structure of the plant C_2S_2-type PSII–LHCII super-complex at a resolution of 3.2 Å (pdb reference 3JCU). **a** View from the stromal side along membrane normal. **b** Side view along membrane plane. The dashed line indicates estimated interfacial regions between the two monomers. The major components are shown as cartoon and stick models in different colors. The 12 small intrinsic subunits are shown as yellow sphere models. **c** A view from the stromal side of the membrane of the plant PSI-LHCI super-complex at a resolution of 2.8 Å (pdb reference 4Y28). The light-harvesting complexes Lhca 1–4 are colored in blue, the PsaH, PsaG and PsaK subunits are colored red. The three iron-sulfur clusters can be distinguished as yellow and red clusters in the middle of the complex

Despite their diversity, all antennae have to operate within the thermodynamic and kinetic constraints dictated by the RCs. Typically, within 100 ps after the initial photon capture by the antenna, very efficient electron transfer occurs from the primary donor, which is P680 in PSII (corresponding to the absorption maximum of the chlorophyll a molecules at 680 nm) to the primary acceptor. To energetically allow excitation energy transfer, the excited state energies of the antenna pigments should be the same or higher than that of the donor, which means—in case of PSII —the absorption maximum of most pigments should preferably lie below 680 nm.

The energy required for charge separation in PSI is somewhat lower, with P700 being the primary donor (the absorption maximum of the chlorophyll a molecules is 700 nm due to the different protein-pigment environment). The antenna complexes transfer light-induced excitons created by the photosynthetically active radiation (PAR, 400–700 nm) to both reaction centers, PSII and PSI [11].

3.1.1.1   Excitation Energy Transfer in the Light-Harvesting Antennae

A fundamental physical process which is used in natural and synthetic light-harvesting to funnel light from any point in the antenna to a 'trap'—the reaction center - is electronic excitation transfer (EET) from an excited molecule (or atom) to another. A successful and widely employed quantitative method for calculating rates of the EET between donor (D) and acceptor (A) pairs was developed by Förster [12]. Theodor Förster connected the energy released by the deexcitation of D and the synchronously uptake by A to the emission and absorption line shapes, respectively. The Förster resonance energy transfer (FRET) is based on electric dipole-dipole interactions between chromophores with a rate, which scales with R^{-6}, where R is the center-to-center distance between the interacting chromophores. The relative orientations of the pigments and the overlap of their fluorescence and absorption spectra determine the transfer rate. Furthermore, the product of the dipole strengths of the corresponding electronic transitions and the overlap of the energy levels is of particular importance. The energy transfer efficiency by the dipole-dipole mechanism and the rate constant can be related to the actual separation R_{DA} of excited donor (*D) and A:

Rate constant for any separation:

$$k_{EET} \propto k_D \left(\frac{R^0_{DA}}{R_{DA}} \right)^6 = \frac{1}{\tau_D} \left(\frac{R^0_{DA}}{R_{DA}} \right)^6 \qquad (3.1)$$

Efficiency for any separation:

$$\phi_{EET} \propto \left(\frac{R^0_{DA}}{R_{DA}} \right)^6 \qquad (3.2)$$

Here, τ_D corresponds to the experimental lifetime of *D, R_{DA} is the actual separation between the centers of *D and A, R^0_{DA} is the critical separation distance between *D and A and ϕ_{ET} is the efficiency for the energy transfer. Therefore, for $R_{DA} < R^0_{DA}$, energy transfer predominates, whereas when $R_{DA} > R^0_{DA}$ the deactivation of *D dominates. When $R_{DA} = R^0_{DA}$, the rate of energy transfer equals the rate of deactivation.

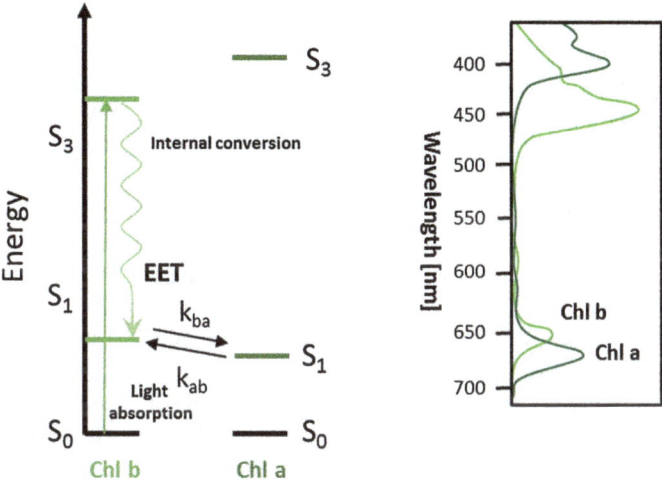

Fig. 3.2 Simplified energy-level diagrams for the chlorophylls a and b (Chl a and Chl b), with their main energy levels corresponding to the absorption peaks. After Chl b is excited by "blue" light which promotes the molecules from the ground state (S_0) to the third excited state (S_3). A rapid relaxation takes place to its first excited state (S_1) from which EET to Chl a follows with a rate k_{ba}. The reverse rate k_{ab} is slower due to the transfer being energetically uphill

Given an environmentally determined refractive index of 1.5, the average transfer rate between two isoenergetic Chl a molecules at R = 1.5 nm with a random orientation is $0.7 - 0.8$ ps^{-1} (van Amerongen et al. [12]). For the energy transfer from a Chl b molecule with an absorption maximum at 650 nm (corresponding to an energy of 3.060×10^{-19} J) to a Chl a molecule with an absorption maximum at 675 nm (2.947×10^{-19} J), the rate k_{ba} is ~ 0.2 ps^{-1} (Fig. 3.2) [11]. The reverse rate k_{ab} can be calculated from $k_{ba}/k_{ab} = \exp(-\Delta/k_b T)$ with $\Delta E = 3.060 \times 10^{-19}$ J -2.947×10^{-19} J. At room temperature (T = 293 K), the reverse rate is about a factor ~ 16 smaller. Therefore, the possibility of uphill energy transfer exists, but it can be considerably slower.

A large part of the EET in photosynthesis proceeds on a timescale below 1 ps where the application of FRET theory is under discussion. The excitation energy is not localized on either the donor or acceptor molecule, but the strong interaction between the pigments results in new excitonic energy levels which are shared between the interacting molecules in a way that they sometimes even result in the formation of one 'super-molecule' [13]. The relative orientation of the pigments influences the dipole strengths for corresponding transitions, which are well-known e.g., for the light-harvesting complexes of purple bacteria (LH1 and LH2). Here, the main absorption bands lie substantially lower in energy than those of the non-interacting pigments. The excitations can also coherently oscillate for a short amount of time between the pigments. The dynamics of this transfer can be described by Redfield theory, considering also interactions with the environment, particularly, vibrations. Strong interactions are generally responsible for transitions between exciton levels, which are caused by the unique design of light-harvesting antennae: the concentration of chromophores (e.g., chlorophylls) can be as high as 0.6 M [14], which is likely to be the key in the design optimization. The inter-chromophore separations are therefore close: in the order of 10 Å center-to-center distance for the nearest neighbours. Since electronic coupling scales steeply with distance, the chromophores in the antenna complex possess strong electronic interactions. Examples of excitons have been discovered in ensemble-averaged linear and non-linear spectroscopy experiments [15, 16] as well as with single-molecule spectroscopy of LH2 complexes from purple photosynthetic bacteria, which have provided strong evidence for the delocalized excited states immediately after the excitation of a single LH2 [17]. Breakthrough experiments using two-dimensional electronic spectroscopy (2DES) have shown evidence of long-lived oscillatory features in the two-dimensional spectra of several light-harvesting complexes. This oscillatory behaviour had been first interpreted as a signature of quantum coherent evolution of superpositions of electronic states, but they were later on attributed to vibrational coupling. Nevertheless, a significant amount of theoretical research aiming at understanding how electronic coherence and in particular, quantum dynamics, can be sustained in the complex biological environment was prompted by this discovery [14]. These theoretical studies investigated whether coherences influence energy transfer dynamics, something which is not detected by the 2DES experiments. A few potential mechanisms supporting coherent dynamics have been identified e.g., weak electron-vibration coupling, spatially correlated environmental fluctuations at different chromophores [18] and a slowly relaxing vibrational environment [19]. Currently, 2DES experi-ments are used to contribute to the understanding of exciton dynamics governed by the balance between the interchromophore coupling (considered small in Förster) and the chromophore-vibrations coupling (considered small in Redfield) [20].

The experimental findings have furthermore raised the question whether coherent dynamics are significant for optimizing photosynthetic light harvesting. It has been recognized that delocalized donor and acceptors states may enhance transport in an incoherent transfer scenario [21], but it is yet unclear whether exciton delocalization between acceptors and donors and the consequent coherent dynamics of excitations provide a significant advantage for light harvesting—despite that it is clear that the mechanism of energy transfer is modified.

3.1.2 Reaction Center Structures and Cofactor Arrangement

Structural models of the photosynthetic RCs, i.e., the ones of purple bacteria and cyanobacterial PSII (Type II family), and eukaryotic and prokaryotic PSI (Type I family) show a similar motif in the arrangement of cofactors involved in the electron transfer reactions: they are organized in two chains ('branches'), arranged along a pseudo-C2 symmetry axis and are located at the interface of the two main

RC subunits. Even though, the 'overall' structural arrangement of the cofactors involved in the primary charge separation reactions are similar in PSI and PSII, there are significant differences in the molecular mechanisms of such reactions: in PSII, only one of the two electron transfer branches is used with a high efficiency ('A branch'), which is known as monodirectional or asymmetric electron transfer (ET). In contrary, both ET branches are used in PSI with a comparable efficiency, described by the terms bidirectional or symmetrical electron transfer. Another relevant difference between the two photosystems is linked to their catalytic activity i.e., the redox potential of the chemical processes involved in the ET reactions. The donor side in PSII is significantly oxidizing ($E° \sim$ +0.8 V) with the intermediate $P680^+$ being estimated as one of the most oxidizing species in nature ($E° \sim$ +1.2 V) [5]. The donor side in PSI is much less oxidizing with $E° \sim$ +0.5 V, but its acceptor side is much more reducing ($E° \sim$ −0.5 V) than the one of PSII ($E° \sim$ 0 V). In both cases, an overall potential difference between the donor and the acceptor side of $\sim$1 V is established, representing about 60% of the energy delivered by a photon corresponding to the lowest excited singlet state transition ($\sim$+1.7 eV). Considering that ΔE between the special pair and the first electron acceptor in PSII (a pheophytin molecule) is about 1.7 V, the thermodynamic efficiency of the primary photochemistry events is about 90%. Energy losses are condoned in order to stabilize primary charge separation events and foster unidirectional electron transfer.

3.1.2.1 Photosystem I

Photosystem I is a membrane chlorophyll-protein complex catalyzing the light-dependent oxidation of plastocyanin (or cytochrome c_6) in the lumen and the reduction of ferredoxin in the stroma. Since the oxidation and reduction of the redox partners occur on opposite sides of the membrane, PSI has to provide an efficient electron transfer across the membrane. Photosynthetic organisms contain the complex in different, but evolutionarily related forms. The most basic structure in all complexes is a symmetric or pseudosymmetric assembly of one or two chlorophyll-containing proteins, which are covalently bound through iron-sulfur clusters serving as electron acceptors [1].

The electron density map of trimeric PSI complexes from the cyanobacterium *T. elongatus* with a resolution of 2.5 Å [22] allowed an accurate model for the protein architecture: PSI of cyanobacteria contains 12 protein subunits bound with four light-harvesting proteins comprising the LHCI antenna complex. The structure of the plant PSI-LHCI supercomplex was recently determined at 2.8 Å resolution. It contains 214 prosthetic groups, including 156 chlorophylls (nine are assigned as chlorophyll b), 32 carotenoids, three [4Fe-4S] clusters, two phylloquinones and 14 lipid molecules [23]. The two transmembrane subunits PsaA and PsaB form a C_2 symmetric heterodimeric core complex and contain most of the cofactors: the pairs of Chl molecules forming the reaction center are arranged in two symmetric branches A and B, which are bound to PsaA and PsaB subunits, respectively.

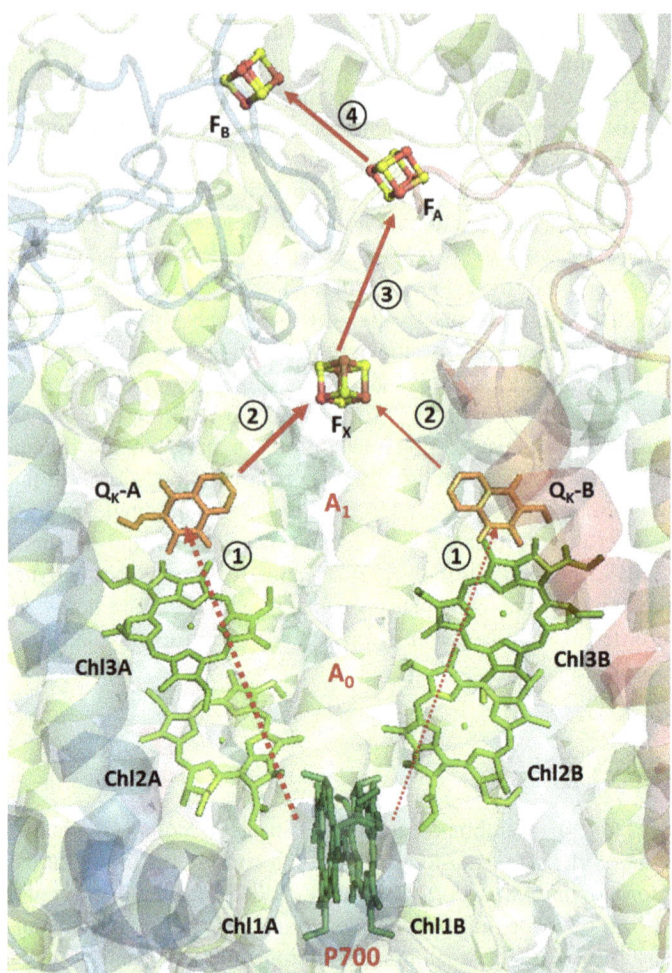

Fig. 3.3 Arrangement of the electron transfer cofactors in PSI from *Thermosynechococcus elongatus* (PDB file 1JB0, Jordan et al. [22]) and a scheme of the electron transfer pathway upon illumination. The numbers represent the order of electron transfer steps after charge separation. Step 1 represents both, the charge separation and the first rapid stabilization step, leading to the formation of $P700^{+}A_1^{-}$ (see text). The photoinduced electron transfer is bidirectional, proceeding through both the A and B branches of cofactors

The electron transfer chain of PSI (Fig. 3.3) consists of two chlorophylls forming P700 (Chl1A/Chl1B or alternatively called eC-A1/eC-B1), two pairs of chlorophylls designated as A_0 (Chl2A/Chl3A and Chl2B/Chl3B, which are also called eC-B2/eC-A3 and eC-A2/eC-B3, respectively), two phylloquinones named as A_1 (designated as Q_K-A/Q_K-B) and the iron-sulfur clusters F_X, F_A and F_B [24].

Following light excitation, the $P700^+A_1^-$ state forms in ~ 50 ps (1) and electron transfer from A_1^- to F_X proceeds biphasically with time constants of 10–25 and 260–340 ns [25] (2)). These two lifetimes are due to the electron transfer down the B and A branches, respectively. While the exact fractional utilization of the two branches appears to be species dependent, optical spectroscopy measurements have recently demonstrated that electron transfer in PSI occurs through both branches of the redox cofactors from P700 to F_X [26]: in cyanobacterial PSI, 66–80% of the electron transfer is contributed by the A-branch and 20–34% is contributed by the B-branch. The reason for the asymmetry and factors influencing it are not quite clear yet. Additionally, the exact kinetics of the primary charge separation events and the nature of the primary electron donor and acceptor in PSI remain contro-versial [27]. Forward electron transfer from F_X^- to F_A and F_B also occurs on a nanosecond timescale at room temperature [29]. In isolated cyanobacterial PSI particles at room temperature, the reduced terminal cluster $F_{A/B}^-$ recombines with $P700^+$ in ~ 100 ms, although the distance between P700 and F_A or F_B is greater than 17 Å, which is too large for a direct electron transfer from $F_{A/B}^-$ to $P700^+$ to occur on this timescale [28]. It is widely accepted that the recombination rather occurs via the repopulation of A_1^-. This notion is supported by the fact that changing the quinone in the binding site modifies the recombination rate.

3.1.2.2 Photosystem II

Photosystem II is a light-driven water-plastoquinone oxidoreductase, carrying out the key reaction of oxygenic photosynthesis: the light-driven oxidation of water [29, 30]. It is a large, multisubunit trans-membrane protein complex found in the photosynthetic membranes of cyanobacteria and photosynthetic eukaryotes and is composed of 17 transmembrane subunits, three extrinsic proteins and several cofactors with a total molecular weight of 350 kDa [31]. A crystal structure is currently available from cyanobacteria at a resolution of 1.95 Å (Fig. 3.4) [32], whereas for plant PSII, it was only recently that the structure of the spinach pho-tosystem II—LHCII supercomplex was obtained at a resolution of 3.2 Å [33]. For a long time, only a three-dimensional model of the supercomplex existed.

The primary processes of the photosynthetic light reactions are initiated by the photoexcitation of the pigments P_{680}, which are formed by the chlorophyll a pair, P_{D1} and P_{D2} and the chlorophylls Chl_{D1} and Chl_{D2} in the PSII reaction center. Within picoseconds, an initial charge separation reaction occurs, involving the formation of $P_{D1}^+P_{D2}^-$, $P_{D1}^+Chl_{D1}^-$ and $Chl_{D1}^+Pheo_{D1}^-$ (with P, Chl and Pheo standing for pair of chlorophylls, monomeric chlorophyll and pheophytin, respectively, Fig. 3.5a [29, 34]), leading to the formation of the charge pair $P_{D1}^+Pheo_{D1}^-$ (1), which in turn reduces another subsequent electron acceptor, the plastoquinone Q_A (2). The electron hole at P_{D1}^+ is able to abstract electrons on the time scale of microseconds from an active tyrosine residue Tyr_Z (3), which in turn reduces the OEC located at the luminal (donor) side (4). Q_A is located close to the stromal

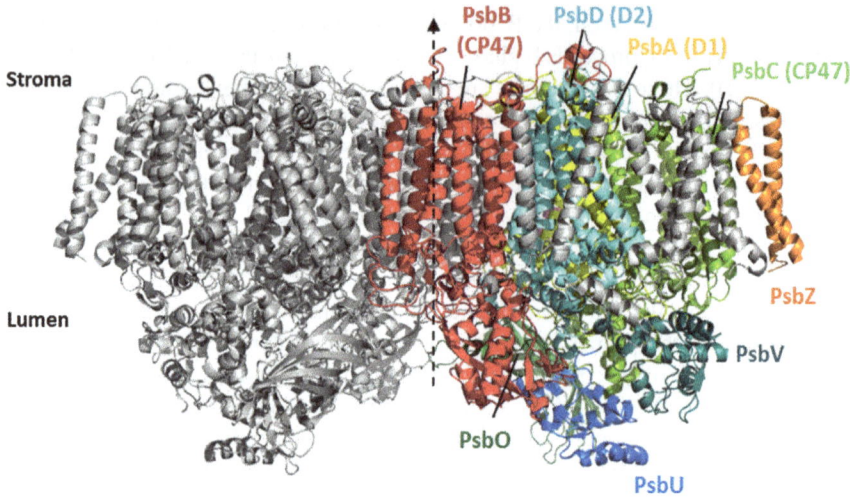

Fig. 3.4 X-ray crystal structure of the PSII dimer from T. vulcanus at a resolution of 1.95 Å, indicating the position of the extrinsic polypeptides, the intrinsic subunits PsbA − D and the small accessory subunit PsbZ (pdb reference 4UB6, Suga et al. [32])

(acceptor) side of the protein and reduces in approximately 0.2–0.4 ms the terminal electron acceptor Q_B (5), which is, once doubly reduced (0.8 ms), released into a membrane quinone pool [34].

The catalytic center of the water oxidation catalyst is composed of four Mn-ions and one Ca^{2+}-ion coordinated by μ-oxo bridges and amino acid residues. Four successive light reactions drive the four successive oxidations between the S_i and S_{i+1} states [35]. Upon the formation of the S_4 state, one molecule of oxygen is produced and released, regenerating the S_0 state. Although the Kok cycle [36] rationalizes already the period-four oscillation of flash-induced oxygen evolution and succeeds in formally describing the function of the enzyme, it does not contain information about the chemical nature of the involved transient states nor the precise mechanisms underlying the interconversion of the S states and the absolute oxidation states of the individual Mn ions in the different S states [37]. Recent computational analyses support the widely accepted hypothesis of the 'high-valence scheme', where the Mn oxidation states are assigned as III, IV, IV, IV in the S_2 state [38]. Figure 3.5b provides an overview of the current view of the extended Kok cycle, indicating the oxidation and proton release events at each transition upon photon absorption by P_{680}.

(a)

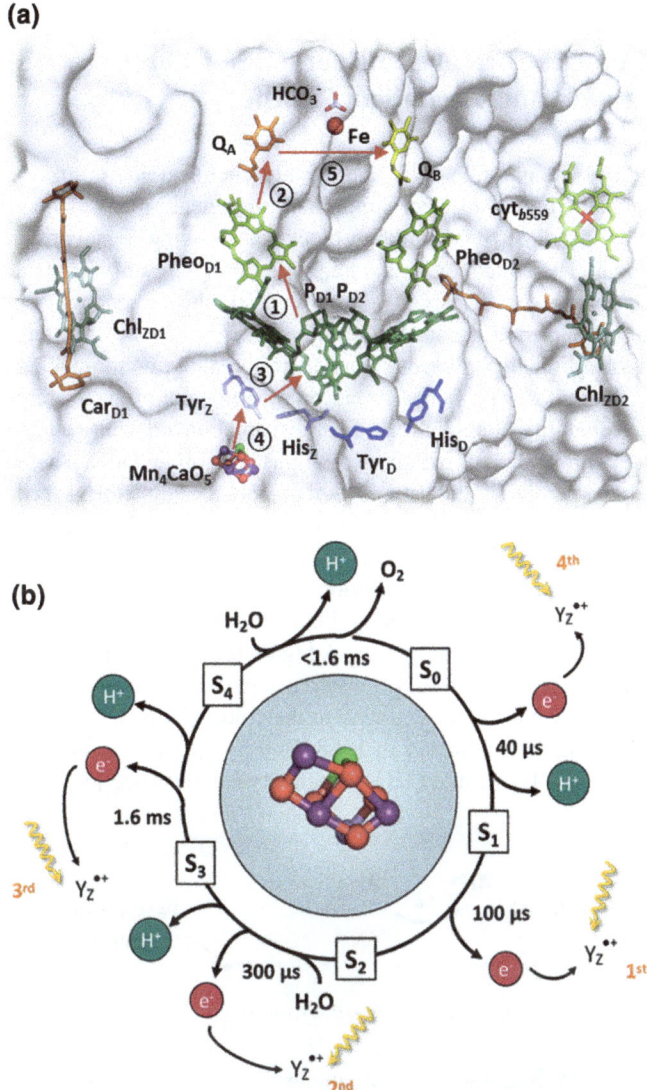

(b)

Fig. 3.5 **a** Schematic representation of the arrangement of cofactors involved in the electron transfer chain in PSII according to the crystal structure (pdb reference 4UB6). The numbers represent the order of electron transfer steps after charge separation. Step 1 represents both, the charge separation and the first rapid stabilization step, leading to the formation of $P_{D1}^+Pheo_{D1}^-$ (see text). **b** Model of the S-state cycle firstly proposed by Kok et al. [36], showing the electron and proton release in the individual S-states (Haumann et al. 2005) and the net oxidation state of the respective S-state (high-valence model, Krewald et al. [38]). A complete cycle requires sequential absorption of four photons, whereas each photon leads within <1 μs to the formation of $Y_Z^{\bullet+}$

3.2   Semiconductor Photochemistry and Principles of Photoelectrocatalysis: Energetics, Charge Separation and Excitation Transfer

With the ability of converting solar energy into chemical energy, oxygenic photosynthesis remains one of the most important biological processes on earth, producing directly or indirectly the building blocks of all living organisms and also of oil, gas and coal. The collection and storage of solar energy in chemical bonds as accomplished by photosynthesis is a highly desirable approach to solving the energy challenge and meeting our future global energy demands. In this term, the development of artificial photosynthetic systems, capable of mimicking this process, is attracting extensive interests. One promising solution to overcome fluctuations in the availability of sunlight is to harvest and convert it into a chemical fuel e.g., H_2, which can be stored, transported and used upon demand. These so-called 'solar fuels' have potential applications as transport fuels, chemical feedstock and as fuels for electricity generation out of daylight hours. The design of such a biomimetic system, however, is not a trivial undertaking, since (i) the solar energy conversion process, storage and distribution should be environmentally benign and protect ecosystems instead of weakening them and (ii) the developed system has to provide a constant and stable energy flux. A highly promising approach to meet this challenge is the photoelectrolysis of water using semiconductors as both, light absorbers and energy converters analogue to the chlorophylls in the photosynthetic reaction centers (Fig. 3.6). Already in 1968, Boddy reported the light-driven oxygen evolution at an n-type rutile (TiO_2) single-crystal electrode [39].

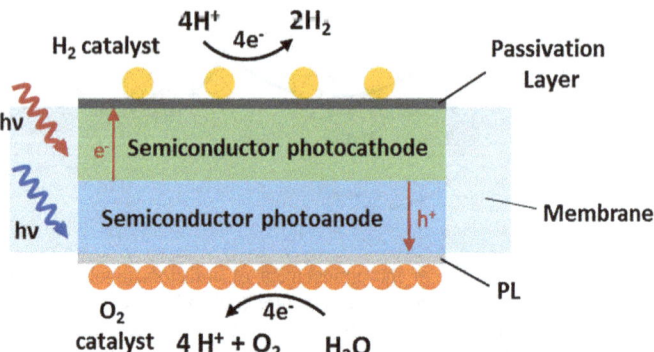

Fig. 3.6 Basic scheme of a monolithic tandem structure device for the four-electron process of water oxidation combined with hydrogen production following the approach of the photosynthetic Z-scheme. The two photoelectrodes possess two different band gaps to absorb different parts of the solar spectrum to generate electrons and holes of a particular energetic level required for the respective half-cell reaction. Deposited electrocatalysts enhance the catalytic reaction. Furthermore, the photoelectrodes are protected against photocorrosion reactions by a passivation layer

This work preceded the highly cited and more famous Nature article by Fujishima and Honda [40], in which the team reported the photoelectrochemical de-composition of water at an n-type TiO_2 photoanode illuminated with ultraviolet illumination in a photoelectrochemical cell with a platinum cathode [40]. Further on, Gerischer [41] proposed the use of photoelectrochemical cells for the conversion of solar energy to electricity using reversible redox systems as the electrolyte [41]. More than 40 years later, the approach to solving the 'holy grail' of artificial water splitting has been focused on investigating new materials for both, the anodic and cathodic processes and integrating configurations which utilize photovoltaic cell junctions in order to increase the obtainable voltage for a single or dual band gap device. The goal is the design of an efficiently operating photoelectrolysis cell, coupling unassisted water oxidation with hydrogen production. This approach requires semiconductor materials which support rapid charge transfer at a semiconductor/ liquid interface, exhibit long-term stability and which can efficiently harvest a large portion of the solar spectrum [42].

The free energy change corresponding to the conversion of one molecule of H_2O to H_2 and $\frac{1}{2}$ O_2 is $\Delta G = 237.2$ kJ/mol or according to the Nernst equation to $\Delta E^0 =$ 1.23 V per electron transferred. Therefore, two minimum requirements result for the photoelectrocatalysis of water at a single semiconductor photoelectrode without applied voltage: firstly, the free energy stored in photogenerated electron-hole pairs in the semiconductor must be large enough to exceed the energy separation between the H^+/H_2 and O_2/H_2O redox levels (1.23 V at 298 K). Secondly, the free energy levels of the holes and electrons must span the redox energy levels in order that proton reduction and water oxidation proceed i.e., the conduction band energy of the semiconductor photocathode must lie above the H^+/H_2 redox level and the valence band of the semiconductor photoanode must lie below the O_2/H_2O level [43]. The semiconductors must therefore absorb radiant light with photon energies of >1.23 eV (equal to wavelengths of ~ 1000 nm and shorter) and the process has to generate two electron-hole pairs per molecule of H_2 (2 × 1.23 eV = 2.46 eV, hydrogen evolution reaction, HER) or four electron-hole pairs per molecule of O_2 (4 × 1.23 eV = 4.92 eV, oxygen evolution reaction, OER):

$$2H^+ + 2e^- \rightarrow H_2 \qquad \text{(HER)} \qquad (3.3)$$

$$H_2O \rightarrow 2e^- + \frac{1}{2}\,O_2 + 2H^+ \qquad \text{(OER)} \qquad (3.4)$$

$$H_2O \rightarrow \frac{1}{2}\,O_2 + H_2 \qquad \Delta G = +237.2 \text{ kJ/mol} \qquad (3.5)$$

Substantial effort was invested in identifying single semiconducting materials, capable of reducing hydrogen and at the same time, generating sufficient external chemical potential (free energy, $\Delta\mu_{ex}$) with the absorbed photons to overcome the 1.23 V required to split water. These single band gap devices have, however, limited values of photoelectrochemical solar to electrical energy conversion efficiencies: beside the capability of absorbing substantial solar light harvesting, they also need to provide adequate stability in harsh aqueous electrolytes and appropriate

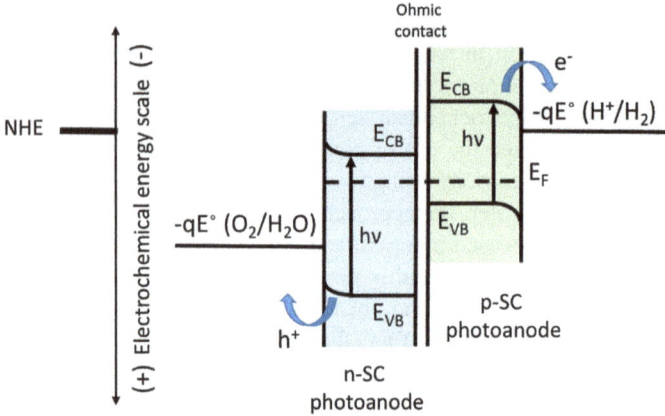

Fig. 3.7 Energy scheme of a photoelectrochemical dual-absorber tandem cell with an n-type semiconductor photoanode and a p-type semiconductor photocathode. The two materials possess two different band gaps, absorbing photons of two different wavelengths. The absorption of a photon (hv) by a semiconductor with a band gap E_g created an electron-hole pair which is separated by the space charge layer to generate a photopotential (as indicated by the Fermi level, E_F, broken line). Combining two absorbers, increases the free energy ($\Delta\mu_{ex}$) available, which must be greater than the energy needed for water splitting plus the overpotential losses at both, anode and cathode, η_{ox} and η_{red}

conduction and valence band levels to straddle the water reduction and oxidation potentials. Since the water splitting reaction requires two half-cell reactions, two or more band gap configurations are investigated (multiple band gap photoelectrochemical cells, MPEC), following the photosynthetic Z scheme with two light-absorbing 'centers'. A simple two-photoelectrode approach uses an n-type semiconductor photoanode and a p-type semiconductor photocathode (Fig. 3.7).

The two semiconductor materials have two different band gap energies, E_{g1} and E_{g2}, respectively, with $E_{g1} > E_{g2}$. In the two photoelectrodes, each absorbed photon creates an excited electron-hole pair, four photons (two in each absorber) have to be absorbed to create one molecule of H_2. Weber and Dignam [44] evaluated the potential solar to hydrogen efficiency of such a device, η_{STH}, compared to placing two cells of the same band gap energy side by side [44]. The evaluation of the side-by-side approach gave an upper limit of 16.6% with $E_{g1} = E_{g2} = 1.4$ eV which represents a decent increase over the predicted single absorber approach of 11.6%. The integrated tandem approach, where the cells were placed on top of each other (see Fig. 3.6) with $E_{g1} = 1.8$ and $E_{g2} = 1.15$ eV, respectively, resulted in 22%. Recent calculations by Prévot and Sivula [45] show that with optimum values of $E_{g1} = 1.89$ and $E_{g2} = 1.34$ eV and assumed losses, an efficiency of $\eta_{STH} = 21.6\%$ can be achieved in the tandem configuration [45] (also compare Chap. 9). The achievement of this solar-to-hydrogen efficiency depends largely on the identification of materials with the correct band gap energies and band positions suitable for the water oxidation and reduction reaction. Therefore, in order to design an efficiently operating solar-water splitting cell, it is also crucial to understand the fundamental photoelectrochemical properties of a semiconductor.

3.2.1 Photoelectrochemistry of Semiconductors

The semiconductor/electrolyte energetics have been described in detail by Gerischer [46]. When a semiconductor with a Fermi level (E_F) is brought in contact with an electrolyte whose electrochemical (redox) potential is E^0, an equilibrium of the two phases is established by transfer of electrons across the surface which leads to a potential barrier for the further flow of charge carriers [46, 47]. The semiconductor electronic bands near the surface bend due to the depletion of majority charge carriers near the surface. The magnitude of the band bending (also potential barrier, E_B) is equal to the difference in electrochemical potentials (Fermi levels) of the two phases, semiconductor and electrolyte, before coming in contact and can be described by:

$$E_B = \left|\left(E_F - E^0\right)\right| \tag{3.6}$$

The situation is analogous to a semiconductor/ metal (solid state) Schottky barrier contact. For an n-type semiconductor, the energy diagram before and after contacting a metal or electrolyte are given in Fig. 3.8. The equilibrium process in (B) gives rise to a space charge layer or depletion layer inside the semiconductor which is depleted of charge carriers. The width of this layer is usually on the order of a few thousand Ångström units and hereby, significantly larger than the Helmholtz double layer caused by smaller carrier densities inside the semiconductor. The potential across the Helmholtz layer is mostly independent of an externally applied potential, whereas most of the potential drops across the semiconductor depletion layer.

The application of an external voltage or light excitation modifies the carrier distribution in the space charge region and therefore varies the Fermi level position and band bending. Absorbed light energy greater than the band gap (E_g) generates excess charge carriers. The electron-hole pairs in the space charge region are separated: the majority carriers move towards the bulk and the minority carriers (electrons and holes respectively for n-type semiconductors) move towards the surface of the semiconductor (C). The net effect is that the potential barrier and band bending are decreased. When the light intensity is sufficiently high, band bending is eliminated and the flat band situation is reached: at this point, the electrochemical potentials of semiconductor and electrolyte are no longer in equilibrium and tend to reach their original positions before contact. This difference in the Fermi levels can be measured as the open circuit potential, V_{OC} (in Fig. 3.8c described as V_P). The maximum photopotential (V_{pmax}) for a given semiconductor/ electrolyte junction upon illumination is therefore given as:

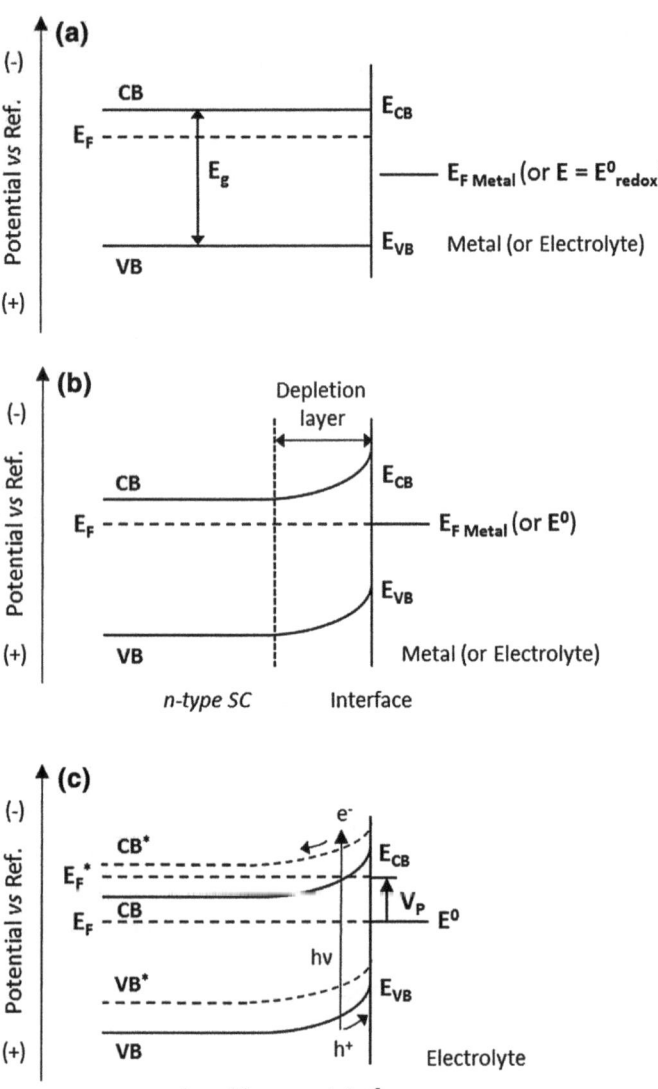

Fig. 3.8 Energetics of the semiconductor/electrolyte interface. **a** n-type semiconductor before contacting a metal or electrolyte (flat band sitation, $E_F = E_{FB}$); **b** n-type semiconductor in charge transfer equilibrium with a metal or electrolyte; **c** n-type semiconductor/ electrolyte interface under illumination with $h\nu \geq E_g$; light generated holes which move towards the semiconductor surface whereas the electrons move to the bulk. E_{CB} and E_{VB} correspond to conduction and valence band edge; E_F and $E_{F\ Metal}$ are the Fermi levels of the semiconductor and the metal; E_0 or E_{redox} are the standard electrochemical potential of the electrolyte; E_g is the band gap energy and V_P corresponds to the photopotential. According to Aruchamy et al. [47]

$$eV_{P,max} = E_B = \left| (E_F - E^0) \right| \tag{3.7}$$

For efficient energy conversion, it would therefore be desirable to select a redox couple where E^0 lies close to E_{VB} for an n-type semiconductor or E_{CB} for a p-type semiconductor in order to maintain a high E_B/E_g ratio [47].

In a photoelectrolysis cell, the redox couple is fixed by the undergoing chemical reaction, i.e. water oxidation and hydrogen production. Therefore, the interface energetics cannot be tuned by simply selecting an appropriate redox couple. One approach is to directly tune the band edge positions relative to the water oxidation or reduction potentials via the incorporation of fixed dipoles or charges on the semiconductor surface: ions adsorbing onto the semiconductor surface have been shown to modify the band edge positions photoelectrode [48, 49].

3.2.2 Photoelectrode Materials

Essential for the development of efficiently operating photoelectrolysis cells are semiconductor materials, which harvest sunlight efficiently and sustain the respective half-cell reaction under illumination for prolonged periods at sufficient enough rates without substantial kinetic losses. Hereby, efficient light harvesting depends not only on the band gaps, but also on the depth of light penetration through the photoelectrode. The light penetration depth is inversely proportional to the absorption coefficient. Semiconductors with moderate light harvesting properties must have good minority carrier diffusion lengths or a low dopant concentration in order to ensure efficient separation of electrons and holes [43].

The identification of materials which meet these requirements is far from being easy and over the last decades, many semiconductor materials have been studied as photoanodes and photocathodes (Fig. 3.9).

Fig. 3.9 Energy band edge positions (conduction band left bar, valence band right bar) and band gaps of common semiconductors used in photoelectrocatalysis cells in relation to the thermodynamic proton reduction and water oxidation potentials and their idealized short circuit current densities for AM 1.5. Redrawn from Walter et al. [42]

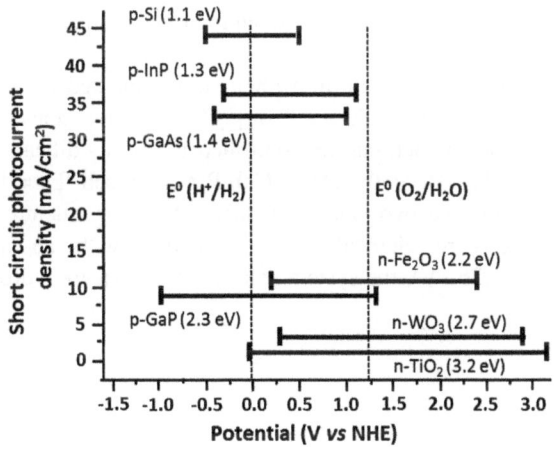

3.2.2.1   Photocathode Materials for Hydrogen Evolution

Photoelectrodes used for hydrogen evolution need to supply a sufficient cathodic current in order to reduce protons to H_2 and they must be stable in aqueous environments. Furthermore, the potential of the conduction band edge has be more negative than the hydrogen redox potential (Fig. 3.7). As described earlier, a semiconductor which is brought in contact with an electrolyte will experience a Fermi Level equilibrium with the electrochemical potential of the liquid by transferring charge across the interface. For a p-type semiconductor, band bending occurs in such a way that photogenerated electrons are driven towards the interface and remaining holes move towards the bulk of the solid. Upon photoexcitation, electrons are injected from the solid into the solution which may protect the surface of the semiconductor from oxidation. Therefore, p-type semiconductors are mostly more stable for HER than n-type semiconductors. One example is p-type GaP with an indirect band gap of 2.26 eV, which is stable for extended periods of time under reducing conditions [42]. p-type GaP can produce H_2 positive of the H^+/H_2 thermodynamic potential [50]. One drawback is that p-GaP possesses small minority-carrier diffusion lengths relative to the absorption depth of visible light in the solid [51]. Another prominent example is InP, which has a band gap of 1.35 eV. A solar-to-hydrogen conversion efficiency of 12% was achieved by Heller and Vadimsky [52], depositing Ru catalysts on the surface of oxidized p-InP [52]. Electrodeposited Rh and Re even resulted in efficiencies of 13.3% and 11.4%, respectively [53]. Furthermore, p-type Si is a small band gap absorber (E_g = 1.12 eV) and an interesting candidate for p/n PEC water splitting cells [54]. It has been demonstrated by multiple groups that planar p-Si photocathodes in combination with a variety of deposited metal catalysts can be used to reduce the voltage required to electrochemically produce H_2 [55, 56]. One obstacle with silicon is its long-term stability: it is stable in acidic solutions, but surface oxidation occurs over extended periods of time. Surface passivation by covalently attaching methyl groups has been demonstrated to improve the stability of the photocathodes [54]. Further materials which have been investigated also include binary oxide photocathodes such as Cu_2O and CuO and chalcopyrite materials such as $CuInS_2$ and $CuGaS_2$ [57].

Currently, the thin-film photovoltaic marked is dominated by II–VI semiconductors such as CdTe and $CdIn_{1-x}Ga_xSe_2$ compounds [42]. These materials possess band gaps which can be controlled by the modification of their composition and processing. Recently, also a GaInP n-p top cell (E_g = 1.78 eV) was used as a light absorber in a two-junction tandem absorber structure with Ge serving as photovoltaic core element. In combination with an n-i-p GaInAs bottom cell (E_g = 1.26 eV), the system resulted in a solar-to-hydrogen efficiency of 14% [58].

3.2.2.2 Photoanodes for Water Oxidation

The key challenge for realizing a biomimetic solar-water splitting device is the solar-driven oxidation of water in analogy to the Mn_4CaO_5 cluster in oxygenic photosynthesis. Many n-type semiconductors have been considered for the water oxidation half-cell reaction, but identifying a material which remains stable in the harsh reaction conditions has been proven to be difficult due to the fast oxidation of the photoanode surface. Furthermore, if the interfacial kinetics of the OER are rate limiting, an OER catalyst needs to be deposited on the electrode. Given the required stability under oxidizing conditions, most of the materials which have been investigated are metal oxides or metal oxide anions (oxo-metalates) in pure, mixed or doped forms [42]. A general feature of these materials is that the valence band consists of O 2p orbitals and the conduction band is formed by the valence orbitals of one more metal. This means that especially in ionic crystals, the potential of the valence band edge stays almost unchanged at 3.0 ± 0.5 V versus NHE for most metal oxides and oxometalates including TiO_2, WO_3, Fe_2O_3 and ZnO. Metal ions can either as bulk matrix or dopant species tune the conduction band position and therefore, also the band gap.

Upon photoexcitation and charge separation of an n-type semiconductor, the minority carriers (holes) in the valence band diffuse to the semiconductor electrolyte interface and oxidize water. The difference between the oxygen-centered valence band (~ 3.0 V) and the water oxidation potential of 1.23 V versus NHE is a major challenge for the development of highly efficient photoanode materials. Much of the excess ~ 1.77 eV absorbed by the metal oxide is wasted by thermal relaxation. A very few semiconductors satisfy both requirements of electronic structure and stability for photoanodes and most examples of functioning photoelectrodes convert sunlight to O_2 at relatively low efficiencies [42].

Several transition metal oxides have been shown to meet some of the requirements for efficient photoelectrochemical water oxidation. In metal oxide photoanodes, the redox-active metal ions include early transition metal oxides such as Ce (IV), Ti(IV), Zr(IV), Nb(V), Ta(V) and d^{10} configuration ions e.g., Zn(II), Ga(III), Ge(IV), Sn(IV) and Sb(V). TiO_2 has been intensively studied following the experiments by Fujishima and Honda, also because of its application in dye-sensitized solar cells [59]. The potential of the TiO_2 conduction band lies slightly above the HER potential (Fig. 3.8) and therefore, electrons in the conduction band do not affect the net reduction of water into H_2 unless the photoelectrode is operated under non-standard conditions (e.g., with a pH gradient between photoanode and photocathode [42]). This limitation can be overcome by using titanates such as $SrTiO_3$ and $BaTiO_3$—the addition of Sr^{2+} and Ba^{2+} ions result in a perovskite structure and the conduction band edge is moved more negative than NHE.

Several strategies have been used to decrease the band gap of oxides and oxometalates that have a d^0 and d^{10} configuration. Metal ions having a strong polarizing capability which result in a strong metal-oxygen bond which has a substantial covalent character and therefore, produce an oxygen-to-metal charge transfer absorption in the visible region [60]. Furthermore, the potential of the conduction band edge can be made more positive by incorporating transition metals and the introduction of N^{3-} and S^{2-} anions can shift the valence band edge to more negative potentials [61]. High nuclear charges of Mo^{6+} and W^{6+} enable these ions to form oxygen-metal networks giving rise to the yellow colour of oxides and the polyoxometalate clusters of these materials [42]. Tungsten oxide (WO_3) is an n-type semiconductor which has been extensively studied for the use as a photoanode in solar water-splitting [62, 63]. It has an indirect band gap energy of 2.7–2.8 eV and can therefore utilize only a limited amount of visible light. While its valence band maximum can provide sufficient overpotential for photogenerated holes to oxidize water, its conduction band position (+0.3 V versus NHE) does not enable photoexcited electrons to reduce water to hydrogen. Computational studies suggest that under working conditions, its surface is completely covered by oxygen atoms and water oxidation takes place via the hydroperoxide and hydroxyl intermediates. WO_3 is also one of the few oxides which are chemically stable in acidic aqueous media. Furthermore, it is inexpensive and non-toxic [62].

Recently, also many coloured composite oxides composed of soft metal ions such as Bi^{3+} and Pb^{2+} have been synthesized. Monoclinic $BiVO_4$ can absorb a substantial portion of the visible spectrum (E_g = 2.4 eV) and it has a favourable conduction band edge position which is near the thermodynamic H_2 evolution potential, allowing $BiVO_4$ to achieve more than 1 V of photovoltage for water oxidation [64]. These favourable features make it to the most recent exciting development in photoanodes for water-splitting PECs.

Group 7–10 transition metal oxides have been widely used as OER cocatalysts [42], although α-Fe_2O_3 (hematite) is one of the most extensively studied n-type binary oxide for solar-water splitting. The band gap (2.0–2.2 eV) allows for the utilization of a significant portion of visible light; a theoretical maximum of its solar-to-hydrogen efficiency exceeds 12% [65, 66]. Below the absorption edge, two weak peaks are present, at 1.4 and 2.0 eV, which can be attributed to crystal field transitions [67]. This results in an intrinsically high charge recombination rate in hematite with a hole-diffusion length of only 2–4 nm [68].

The key challenge for efficient water oxidation at semiconductor surfaces is the development of a stable photoanode which absorbs visible light. The ideal light absorber would behave like a low-pass colour filter whose extinction coefficient approaches infinity above the band gap of the absorber. These materials possess a pure, rich colour and include materials which are widely used as pigments such as $BiVO_4$. In a direct band gap semiconductor, photon absorption is complete within a surface depth of 100–1000 nm and charges do not need long distances before reaching the solution. Therefore, these materials can be easily made into nanoparticles or thin films for water-splitting devices. An indirect band gap results in a weaker absorption at longer wavelengths and produces an unsaturated colour [42]. Unfortunately, it is not easy to predict based on the crystal structure or putative electronic structure whether a semiconductor's band gap is direct or indirect; e.g., GaN, GaP and GaAs all exist in the zincblende structure, but GaP has an indirect band gap and GaAs and GaN have direct band gaps.

Arguing from a material's perspective, probably many oxides and structured, wide band gap materials have many of the desired characteristics for efficient photoanodic water oxidation. Osterloh [69] provides a recent review on the metal oxides which have been studied for electrochemical water-splitting [69]. To facilitate the search for new, promising materials, efficient screening procedures have been described e.g., by Woodhouse and Parkinson, in which the metal oxide compositions are electrochemically deposited onto electroplates using robotics and are individually screened for water photooxidation activity [70].

Dividing light absorption and charge separation into two photosystems according to the energy needed to drive the respective half-cell reaction seems to be a promising approach taken by nature for the construction of efficient photoelectrochemical devices. After more than 40 years of research, the goal of an efficient and stable tandem cell seems to be within reach. Although, long-term stability and higher solar-to-hydrogen conversion efficiencies are still needed, the recent and rapid developments of new photoelectrode materials along with the continued understanding of the photoelectrochemical processes at the electrodes suggest that the remaining challenges can be overcome in the upcoming decades and efficient global-scale solar energy conversion can be achieved.

References

1. N. Nelson, W. Junge, Structure and energy transfer in photosystems of oxygenic photosynthesis. Ann. Rev. Biochem. **84**, 659–683 (2015)
2. N. Nelson, Evolution of Photosystem I and the control of global enthalpy in an oxidizing world. Photosynth. Res. **116**, 145–151 (2013)
3. A. Ben-Shem, A. Frolow, F. Nelson, Crystal structure of plant Photosystem I. Nature **426** (6967), 630–635 (2003)
4. H.S. Yoon, J.D. Hackett, C. Ciniglia, G. Pinto, D. Bhattacharya, A molecular timeline for the origin of photosynthetic eukaryotes. Mol. Biol. Evol. **21**, 809–818 (2004)
5. S. Caffarri, T. Tibiletti, R.C. Jennings, S. Santabarbara, A comparison between plant Photosystem I and Photosystem II architecture and functioning. Curr. Protein Pept. Sci. **15**, 296–331 (2014)
6. P. Horton, A.V. Ruban, R.G. Walters, Regulation of light harvesting in green plants. Ann. Rev. Plant Physiol. Plant Mol. Biol. **47**, 655–684 (1996)
7. M.E. Hohmann-Marriott, R.E. Blankenship, Evolution of photosynthesis. Ann. Rev. Plant Biol. **62**, 515–548 (2011)
8. B. Green, W.W. Parson, *Light-Harvesting Antennas in Photosynthesis* (Springer Science and Business Media, Dordrecht, 2003)
9. R.E. Blankenship, *Molecular Mechanisms of Photosynthesis*, 2nd edn. (Oxford, Wiley Blackwell, 2014)
10. R. Croce, H. van Amerongen, Natural strategies for photosynthetic light harvesting. Nat. Chem. Biol. **10**, 492–501 (2014)
11. T. Förster, Zwischenmolekulare Energiewanderung und Fluoreszenz. Ann. Phy. **437**(1–2), 55–75 (1948)
12. H. Van Amerongen, R. van Grondelle, L. Valkunas, *Photosynthetic Excitons*, 1st edn. (World Scientific Pub. Co., Inc., Singapore, 2000)

13. F. Fassioli, D. Rayomond, P.C. Arpin, G.D. Scholes, Photosynthetic light harvesting: excitons and coherence. J. R. Soc. Interface **11**(92), 20130901–20130922 (2013)
14. V.I. Novoderezhkin, V.I. Razjivin, Excitation delocalization over the whole core antenna of photosynthetic purple bacteria evidenced by non-linear pump-probe spectroscopy. FEBS Lett. **368**, 370–372 (1995)
15. R. Monshouwer, M. Abrahamsson, F. van Mourik, R. van Grondelle, Superradiance and exciton delocalization in bacterial photosynthetic light-harvesting systems. J. Phys. Chem. B **101**, 7241–7248 (1997)
16. R.J. Cogdell, A. Gall, J. Kohler, The architecture and function of the light-harvesting apparatus of purple bacteria: from single molecules to in vivo membranes. Q. Rev. Biophys. **39**, 227–324 (2006)
17. A.M. Van Oijen, M. Ketelaars, J. Kohler, T.J. Aartsma, J. Schmidt, Unravelling the electronic structure of individual photosynthetic pigment-protein complexes. Science **285**, 400–402 (1999)
18. A. Ishizaki, G.R. Fleming, Theoretical examination of quantum coherence in a photosynthetic system at physiological temperature. Proc. Natl. Acad. Sci. U.S.A. **106**, 17255–17260 (2009)
19. S. Jang, M. Newton, R. Silbey, Multichromophoric Förster resonance energy transfer. Phys. Rev. Lett. **92**(218301), 1–4 (2004)
20. A. Chenu, G.D. Scholes, Coherence in energy transfer and photosynthesis. Annu. Rev. Phys. Chem. **66**, 69–96 (2015)
21. I. Kassal, J. Yuen-Zhou, S. Rahimi-Keshari, Does coherence enhance transport in photosynthesis? J. Phys. Chem. Lett. **4**, 362–367 (2013)
22. P. Jordan, P. Fromme, H.T. Witt, O. Klukas, W. Saenger, N. Krau, Three-dimensional structure of cyanobacterial Photosystem I at 2.5 Å resolution. Nature **411**, 909–917 (2001)
23. Y. Mazor, A. Borovikova, N. Nelson, The structure of plant Photosystem I super-complex at 2.8 Å resolution. eLife 4, e07433–e07451 (2015)
24. M. Mamedov, V. Nadtochenko, A. Semenov, Primary electron transfer processes in photosynthetic reaction centers from oxygenic organisms. Photosynth. Res. **125** (1–2), 51–63 (2015)
25. H. Makita, G. Hastings, Modelling electron transfer in Photosystem I. Biochim. Biophys. Acta **1857**, 723–733 (2016)
26. O.G. Poluektov, L.M. Utschig, Directionality of electron transfer in type I reaction center proteins: high-frequency EPR study of PSI with removed iron-sulfur centers. J. Phys. Chem. B. **119**, 13771–13776 (2015)
27. A.R. Holzwarth, M.G. Muller, J. Niklas, W. Lubitz, Ultrafast transient absorption studies on Photosystem I reaction centers from Chlamydomonas reinhardtii. 2: mutations near the P700 reaction center chlorophylls provide new insights into the nature of the primary electron donor. Biophys. J. **90**, 552–565 (2006)
28. M. Byrdin, S. Santabarbara, F. Gu, W.V. Fairclough, P. Heathcote, K. Redding, F. Rappaport, Assignment of a kinetic component to electron transfer between iron-sulfur clusters FX and FA/B of Photosystem I. Biochim. Biophys. Acta Bioenerg. **1757**, 1529–1538 (2006)
29. T. Cardona, A. Sedoud, N. Cox, A.W. Rutherford, Charge separation in Photosystem II: a comparative and evolutionary overview. Biochim. Biophys. Acta **1817**(1), 26–43 (2012)
30. F. Rappaport, B.A. Diner, Primary photochemistry and energetics leading to the oxidation of the Mn4Ca-cluster and to the evolution of molecular oxygen in Photosystem II. Coordin. Chem. Rev. **252**(3–4), 259–272 (2008)
31. Y. Umena, K. Kawakami, J.R. Shen, N. Kamiya, Crystal structure of oxygen-evolving Photosystem II at a resolution of 1.9 Å. Nature **473**, 55–60 (2011)
32. Suga et al., Native structure of Photosystem II at 1.95 Å resolution viewed by femtosecond X-ray pulses. Nature **517**(7532), 99–103 (2015)
33. X. Wei, X. Su, P. Cao, X. Liu, W. Chang, M. Li, X. Zhang, Z. Liu, Structure of spinach photosystem II-LHCII supercomplex at 3.2 Å resolution. Nature **534**(7605), 69–74 (2016)

34. H. Dau, M. Haumann, The manganese complex of Photosystem II in its reaction cycle—basic framework and possible realization at the atomic level. Coord. Chem. Rev. **252**, 273–295 (2008)
35. P. Joliot, G. Barbieri, R. Chabaud, Un nouveau modèle des centres photochimiques du système II. Photochem. Photobiol. **10**, 309–329 (1969)
36. B. Kok, B. Forbush, M. McGloin, Cooperation of charges in photosynthetic O_2 evolution. I. A linear four step mechanism. Photochem. Photobiol. **11**, 467–475 (1970)
37. N. Cox, D.A. Pantazis, F. Neese, W. Lubitz, Biological. Acc. Chem. Res. **46**, 1588–1596 (2013)
38. V. Krewald, M. Retegan, N. Cox, J. Messinger, W. Lubitz, S. DeBeer, F. Neese, D.A. Pantazis, Metal oxidation states in biological water splitting. Chem. Sci. **6**, 1676–1695 (2015)
39. P.J. Boddy, Oxygen evolution. J. Electrochem. Soc. **115**(2), 199–203 (1968)
40. A. Fujishima, K. Honda, Electrochemical photolysis of water at a semiconductor electrode. Nature **238**, 37–38 (1972)
41. H. Gerischer, Electrochemical photo and solar cells principles and some experiments. J. Electroanal. Chem. Interfacial Electrochem. **58**, 263–274 (1975)
42. M.G. Walter, E.L. Warren, J.R. McKone, S.W. Boettcher, Q. Mi, E.A. Santori, N.S. Lewis, Solar water splitting cells. Chem. Rev. **110**, 6446–6473 (2010)
43. L.M. Peter, K.G.U. Wijayantha, Photoelectrochemical water splitting at semiconductor electrodes: fundamental problems and new perspectives. Chem. Phys. Chem **15**, 1983–1995 (2014)
44. M.F. Weber, M.J. Dignam, Efficiency of splitting water with semiconducting photoelectrodes. J. Electrochem. Soc. **131**, 1258–1265 (1984)
45. M.S. Prévot, K. Sivula, Photoelectrochemical tandem cells for solar water splitting. J. Phys. Chem. C **117**(35), 17879–17893 (2013)
46. H. Gerischer, Semiconductor electrochemistry, in *Physical Chemistry. Vol. IXA: Electrochemistry,* ed. by H. Eyring, D. Henderson and W. Jost (Academic Press, New York 1970), pp. 463–542
47. A. Aruchamy, G. Aravamudan, G.V. Subba Rao, Semiconductor based photoelectrochemical cells for solar energy conversion—An overview. Bull. Mater. Sci. **4**(5), 483–526 (1982)
48. S.S. Kocha, J.A. Turner, Displacement of the bandedges of $GaInP_2$ in aqueous electrolytes induced by surface modification. J. Electrochem. Soc. **142**, 2625–2630 (1995)
49. L.A. Lyon, J.T. Hupp, Energetics of semiconductor electrode/solution interfaces: EQCM evidence for charge-compensating cation adsorption and intercalation during accumulation layer formation in the titanium dioxide/acetonitrile system. J. Phys. Chem. **99**, 15718–15720 (1995)
50. R. Memming, G. Schwandt, Electrochemical properties of gallium phosphide in aqueous solutions. Electrochim. Acta **13**, 1299–1310 (1968)
51. D.E. Aspnes, A.A. Studna, Dielectric functions and optical parameters of Si, Ge, GaP, GaAs, GaSb, InP, InAs, and InSb from 1.5 to 6.0 eV. Phys. Rev. B **27**, 985–1009 (1983)
52. A. Heller, R.G. Vadimsky, Efficient solar to chemical conversion: 12% efficient photoassisted electrolysis in the [p-type InP(Ru)]/HCl-KCl/Pt(Rh) cell. Phys. Rev. Lett. **46**, 1153–1156 (1981)
53. E. Aharon-Shalom, A. Heller, Efficient p-InP (Rh - H alloy) and p-InP (Re H alloy) hydrogen evolving photocathodes. J. Electrochem. Soc. **129**(12), 2865–2866 (1982)
54. T.W. Hamann, N.S. Lewis, Control of the stability, electron-transfer kinetics, and pH-dependent energetics of Si/H_2O interfaces through methyl termination of Si(111) surfaces. J. Phys. Chem. B **110**, 22291–22294 (2006)
55. Y. Nakato, H. Yano, S. Nishiura, T. Ueda, H.J. Tsubomura, Hydrogen photoevolution at p-type silicon electrodes coated with discontinuous metal layers. J. Electroanal. Chem. **228**, 97–108 (1987)
56. R.N. Dominey, N.S. Lewis, J.A. Bruce, D.C. Bookbinder, M.S. Wrighton, Improvement of photoelectrochemical hydrogen generation by surface modification of p-type silicon semiconductor photocathodes. J. Am. Chem. Soc. **104**, 467–482 (1982)

57. D. Kang, T.W. Kim, S.R. Kubota, A.C. Cardiel, H.G. Cha, K.-S. Choi, Electrochemical synthesis of photoelectrodes and catalysts for use in solar water splitting. Chem. Rev. **115**, 12839–12887 (2015)
58. M.M. May, H.-J. Lewerenz, D. Lackner, F. Dimroth, T. Hannappel, Efficient direct solar-to-hydrogen conversion by in situ interface transformation of a tandem structure. Nat. Comm. **6**, 8272–8286 (2015)
59. M. Grätzel, Photoelectrochemical cells. Nature **414**, 338–344 (2010)
60. P.M. Woodward, H. Mizoguchi, Y.-I. Kim, M.W. Stoltzfus, The electronic structure of oxides in metal oxides: chemistry and applications, in *Metal Oxides: Chemistry and Applications*, ed. by J.L.G. Fierro (CRC Press, New York, 2006), pp. 133–193
61. A. Kudo, Y. Miseki, Heterogeneous photocatalyst materials for water splitting. Chem. Soc. Rev. **38**(1), 253–278 (2009)
62. D. Kang, T.W. Kim, S.R. Kubota, A.C. Cardiel, H.G. Cha, K.-S. Choi, Electrochemical synthesis of photoelectrodes and catalysts for use in solar water splitting. Chem. Rev. **115**, 12839–12887 (2015)
63. X. Liu, F. Wang, Q. Wang, Nanostructure-based WO_3 photoanodes for photoelectrochemical water-splitting. Phys. Chem. Chem. Phys. **14**, 7818–7894 (2012)
64. F.F. Abdi, L. Han, A.H.M. Smets, M. Zeman, B. Dam, R. van de Kroel, Efficient water-splitting by enhanced charge separation in a bismuth vanadate-silicon tandem photoelectrode. Nat. Comm. **4**, 2195–2202 (2013)
65. R. Van de Kroel, Y. Liang, J. Schoonmann, Solar hydrogen production with nanostructured metal oxides. J. Mater. Chem. **18**, 2311–2320 (2008)
66. K. Sivula, F. Le Formal, M. Grätzel, Solar water splitting: progress using hematite α-Fe_2O_3 photoelectrodes. Chem. Sus. Chem. **4**(4), 432–449 (2011)
67. L.A. Marusak, R. Messier, W.B. White, Optical absorption spectrum of hematite, α-Fe_2O_3 near IR to UV. J. Phys. Chem. Solids **41**(9), 981–984 (1980)
68. J.H. Kennedy, K.W. Frese, Photooxidation of Water at α-Fe2O3 Electrodes. J. Electrochem. Soc. **125**, 709–714 (1978)
69. F.E. Osterloh, Inorganic materials as catalysts for photochemical splitting of water. Chem. Mater. **20**, 35–54 (2008)
70. M. Woodhouse, B.A. Parkinson, Combinatorial approaches for the identification and optimization of oxide semiconductors for efficient solar photoelectrolysis. Chem. Soc. Rev. **38** (1), 197–210 (2008)

Chapter 4
Electron Transfer

4.1 Electron Transfer, Bioenergetics and Redox Potentials of Cofactors in Photosystem II

Electron flow through proteins and protein assemblies of the photosynthetic reaction centres occurs between metal centres or other redox cofactors which are separated by relatively large distances, often in the 10–20 Å range. In order to separate an electron and hole over 25 Å across a membrane (Fig. 4.1), PSII utilizes multi-step charge hopping on many different time scales aiming at loosing minimal amounts of free energy [1, 2].

The starting point for understanding single-step electron transfer (ET) reactions in chemistry and biology is the semi-classical theory of electron transfer reactions, formulated i.a. by Marcus [3, 4] and Levich and Dogonadze [5], providing the theoretical underpinning for numerous experimental investigations:

$$k_{ET} = \sqrt{\frac{4\pi^3}{h^2 \lambda k_B T}} H_{AD}^2 \exp\left(-\frac{(\Delta G^\circ + \lambda)^2}{4\lambda k_B T}\right) \tag{4.1}$$

The theory expresses the specific ET rate between two weakly interacting redox centres, a donor (D) and an acceptor (A), at a fixed distance and orientation in terms of the standard free energy change for the reaction (ΔG°), with a parameter describing the extent of nuclear reorientation and reorganization accompanying the ET (λ) and the electronic coupling strength between reactants and products at the transition-state nuclear configuration (H_{AD}). The probability of electron tunnelling from the donor to an acceptor in the activated complex is described by the exponential factor in which k_B, T and H have their usual meanings (see Fig. 4.2.). H_{AD} is related to the distance between D and A:

© Springer International Publishing AG, part of Springer Nature 2018
K. Brinkert, *Energy Conversion in Natural and Artificial Photosynthesis*,
Springer Series in Chemical Physics 117,
https://doi.org/10.1007/978-3-319-77980-5_4

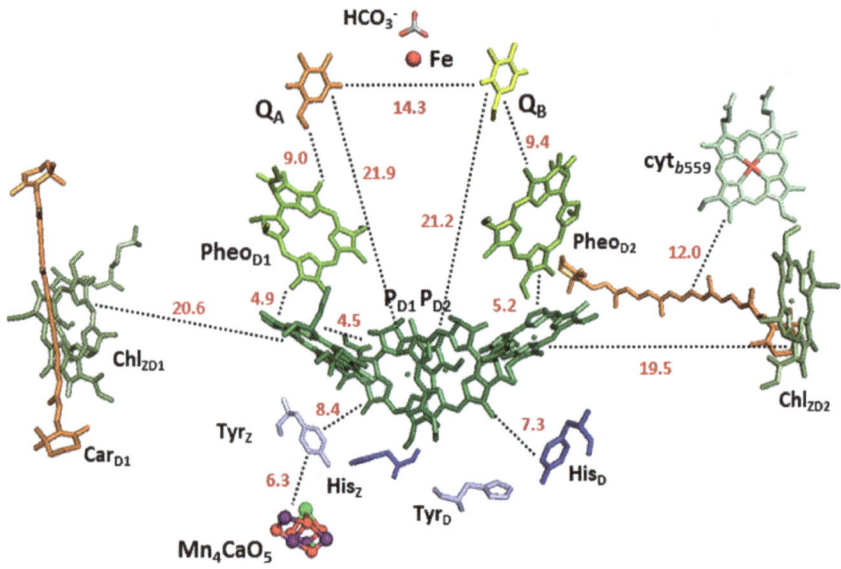

Fig. 4.1 Reaction centre of Photosystem II with indicated cofactor distances [in Å] according to the recent crystal structure (pdb reference 4UB6)

Fig. 4.2 Potential energy well for an electron transfer reaction treated quantum mechanically. R and P are the wells of the reactant and product vibrational energy, respectively. Q* is the coordinate of interest to the system at the transition state. See text for further explanations of the indicated parameters

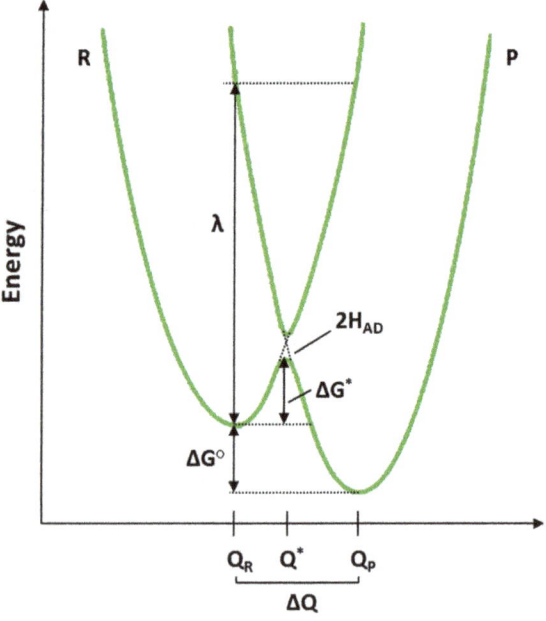

$$H_{AD} = H_{AD}(r_0) \cdot \exp(-0.5\beta(r - r_0)) \qquad (4.2)$$

Here, β is the decay constant for tunnelling, r_0 is the limit of contact (as the sum of Van der Waals radii) and r is the distance between D and A. $H_{AB}(r_0)$ is the electronic coupling matrix element at $r = r_0$. In general, ΔG° and λ depend on the molecular composition and the local environment of D and A, while H_{AB} is a function of the D–A distance and the structure of the intervening medium [2]. A unique feature of homogenous ET reactions is the Gaussian free-energy dependence of the specific rate; in favourable cases, rates are observed to decrease as the driving force increases beyond λ (inverted effect).

Energy saving charge separation reactions in the photosynthetic reaction centres are faster than the energy-wasting recombination processes, which has also been rationalized in terms of the inverted effect. When excitation arrives at the Photosystem II reaction center, a slew of different charge separations occurs in the first few picoseconds, which leads to an electron transfer reaction. This results in the formation of the charge pair $P_{D1}^+Pheo_{D1}^-$ with the cation being mainly localized on P_{D1} [6, 7]. Capturing the initial charge separation process in measurements is complicated and multiphasic due to the difficulty in distinguishing between absorbance changes associated with excited states and radical pair states, which are nearly isoenergetic and coupled with inhomogeneous optical broadening. Femtosecond infrared and visible spectroscopy (T $\geq$ 4 °C) provides quite different numbers for PSII reaction centers isolated from spinach: the formation of $Pheo_{D1}^-$ was detected by Groot et al. (2005) in 0.6–0.8 ps with P_{D1}^+, whereas Holzwarth et al. (2006) found the initial radical pair state containing $Pheo_{D1}^-$ appearing in 3.2 ps, whereas the state $P_{D1}^+Pheo_{D1}^-$ appeared with a life time of 11 ps. Nevertheless, both groups agree that $Pheo_{D1}^-$ appears prior to P_{D1}^+ and they attribute the oxidized donor to $Pheo_{D1}^-$ to Chl_{D1}^+. Chl_{D1} is thought to be the longest wavelength pigment in PSII [8–10], which favors the localization of the exciton on this pigment. The formation of the $Chl_{D1}^+Pheo_{D1}^-$ state is supported as well by Stark spectra, indicating the existence of a $Chl_{D1}^{\delta+}Pheo_{D1}^{\delta-}$ charge transfer state [11]. Electron transfer from $Pheo_{D1}^-$ to Q_A is reported to occur on the time scale of 200–500 ps [12].

Based on the kinetic studies of electron transfer dynamics in PSII (summarized in Fig. 4.3), one can determine the relative free-energy level of various transition states formed during the turnover of PSII, which is essential for the understanding the water-splitting mechanism. Additionally, the measurable midpoint potential of a limited number of titratable electron carriers such as the one of $PheoD_1$ and Q_A (see Sect. 4.3) allow a further completion of the energetic picture: if the free energy change associated with the electron transfer between one of these titratable redox cofactors and another electron carrier is known, one can determine their relative reduction potentials.

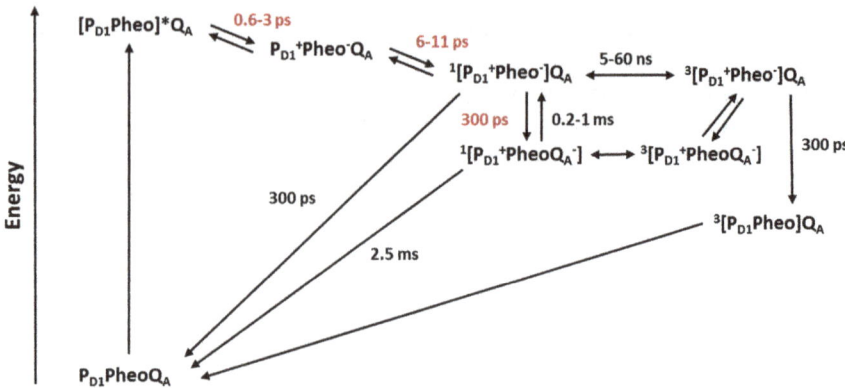

Fig. 4.3 Electron transfer dynamics at the PSII acceptor side. According to Rappaport and Diner (2008)

4.2 Proton-Coupled Electron Transfer in Photosystem II: Kinetics and Thermodynamics

Photosystem II is often evoked as the paradigm of biological proton-coupled electron transfer reactions (PCET), with proton-coupled redox processes occurring during tyrosine oxidation and reduction, water oxidation and quinone reduction at the electron acceptor site. PCET reactions involve the transfer of an electron and a proton, which may be sequential, where either the electron or the proton transfers first, or concerted, where the electron and proton transfer simultaneously [13]. A general PCET reaction can be described in terms of four diabatic electronic states, where *1* and *2* denote the ET state and *a* and *b* denote the PT state:

$$(1a) \quad D_e^- - D_p - H^+ \cdots A_p - A_e$$
$$(1b) \quad D_e^- - D_p \cdots {}^+H - A_p - A_e$$
$$(2a) \quad D_e - D_p - H^+ \cdots A_p - A_e^-$$
$$(2b) \quad D_e - D_p \cdots {}^+H - A_p - A_e^-$$

This model can describe sequential mechanisms, where the proton transfers prior to the electron (i.e., *1a → 1b → 2b*) or the electron transfers prior to the proton (i.e., *1a → 2a → 2b*) as well as concerted mechanisms (i.e., *1a → 2b*). Marcus' treatment of electron transfer (see also previous section) provides a contextual starting point for the PCET reactions [4, 14]:

$$k_{ET} = k_{ET}(0) \exp \left[\frac{-(\lambda + \Delta G^\circ)^2}{4\lambda RT} \right] \tag{4.3}$$

Here, ΔG° is the driving force for the reaction, λ is the energy needed to reorganize the nuclear configuration of the system from the equilibrium configuration of the reactant to the transition state. $k_{ET}(0)$ is the activationless ($\Delta G^\circ = -\lambda$) ET rate constant. The reorganization energy λ is described by Marcus as the sum of energies required to reorganize the bond lengths and angles of the redox cofactor (inner sphere reorganization energy) and the surrounding medium (outer sphere reorganization energy, λ_0). Modifications of (4.3) which consider electronic coupling yield to the previously introduced Marcus-Levich-Hush equation for ET ((4.2), previous section).

In a PCET reaction, every parameter of (4.2) is affected by the proton; the electron movement changes the pK_a of the redox cofactor. In order to predict the kinetics, knowledge of the driving force of the reaction alone is insufficient; the charge redistribution resulting from electron and proton transfer will affect the energy associated with the reorganization of the surrounding environment. Furthermore, the electronic coupling depends on the overlap of both, the electronic and the proton vibrational wavefunctions of the donor and acceptor states. Each will change with the proton vibrational wavefunctions of the donor and acceptor states parametrically [15]. Any proton movement from its initial position will therefore perturb the electronic coupling strength between the donor and acceptor, H_{AD}, ΔG° and λ. Since the process includes the breaking and forming of new bonds, at an extreme level, PCET falls outside the confines of the conventional theory. Therefore, PCET reactions steps are described beyond the discussion of ET, since both, the electron and proton, affect H_{AD} and the Franck-Condon (FC) term (4.2); in an ET, the electron tunnels through the potential barrier from D and A when the medium fluctuates to a configuration where the energies of the donor and acceptor are equal at the surface crossing [15]. Within a PCET reaction, proton tunnelling also occurs, which is furthermore influenced by the fluctuations in the medium. This means that the electron and proton influence each other thermodynamically and kinetically.

PCET has been treated originally by Cukier [16], who added a second dimension into the ET picture to incorporate the proton into the ET description (Fig. 4.4). The electron-proton vibronic surfaces of reactant and product can be calculated in the concerted PCET as functions of two collective solvent coordinates (Z_e (ET) and Z_p (PT), resp.). The reactant state describes the localization of the transferring electron on its donor and the product state corresponds to the localization of the transferring electron on its acceptor. Typically, PCET reactions are nonadiabatic due to the relatively small couplings between the reactant and product vibronic states. The rate constant for a PCET can be described as the sum of the rate constant for nonadiabatic transitions between all pairs of reactant and product vibronic states [17]:

$$k = \frac{2\pi}{\hbar} \sum_\mu^{\{I\}} P_\mu^I \sum_\nu^{\{II\}} |V_{\mu\nu}|^2 (4\pi\lambda_{\mu\nu}k_BT)^{-\frac{1}{2}} \exp\left(\frac{-\left(\Delta G_{\mu\nu}^\circ + \lambda_{\mu\nu}\right)^2}{4\lambda_{\mu\nu}k_BT}\right) \quad (4.4)$$

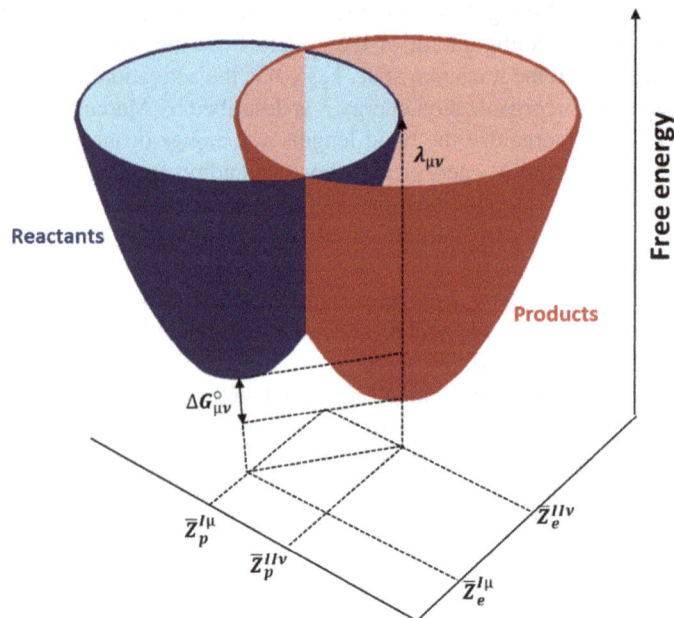

Fig. 4.4 Two-dimensional vibronic free-energy surfaces for reactants I, μ (blue) and products II, μ (red) as functions of two collective solvent coordinates for a PCET reaction. The lowest energy reactant and product free energy surfaces are shown. The free energy difference, $\Delta G^{\circ}_{\mu v}$, and the outer-sphere reorganization energy, $\lambda\mu v$, are indicated. Adapted from Hammes-Schiffer (2006)

Here, the summations are over the reactants' and products' vibronic states associated with the ET states I and II, respectively. P^{I}_{μ} is the Boltzmann probability for the reactant state I μ, $V_{\mu v}$ is the coupling between reactant and product vibronic states, $\Delta G^{\circ}_{\mu v}$ is the free energy of the reaction and $\lambda_{\mu v}$ is the outer-sphere reorganization energy. $\Delta G^{\circ}_{\mu v}$ and $\lambda_{\mu v}$ are depicted in Fig. 4.4. Each channel or vibrational model in the reactant well couples to the product well with a different electronic coupling $V_{\mu v}$. For each vibrational mode, the overall rate is weighted by the Boltzmann factor for the thermal population of the respective channel. Therefore, if the overall driving force of the reaction and the experimental rate constant is known, reorganization energies, electronic couplings and percent contribution to the rate for each reactant-product vibrational channel can be calculated. Although the theory is not predictive a priori, it provides a framework in which PCET reactions can be analysed insightfully e.g., in Photosystem II [18].

4.2.1 The Tyrosyl Radicals in Photosystem II: Why D and Why Z?

Tyrosine is a major player in many important PCET proteins, such as ribonucelotide reductase, galactose oxidase, cytochrome c oxidase and Photosystem II. In the latter one, two redox active tyrosine residues, D1-Tyr161 (Tyr_Z) and D2-Tyr160 (Tyr_D) are located in homologous positions in the D1 and D2 protein core subunits of PSII, respectively (see Fig. 3.5a). Tyr_Z directly participates in the catalytic process of PSII, serving as an electron relay between P680 and the Mn_4CaO_5 cluster, whereas TyrD on the other hand could be considered as an evolutionary relict, which is not directly involved in the catalytic activity of PSII. Nevertheless, TyrD is strictly conserved among all oxygenic species [19].

In the early 1950s, Electron Paramagnetic Resonance (EPR) was introduced to biological materials. Soon later, in 1956, Commer and co-workers investigated a chloroplast solution from a tobacco plant in the EPR cavity for the first time [20]. Two EPR signals were discovered: one was formed immediately when the suspension was exposed to light and the other signal was a narrow radical signal which decayed quickly when the light was turned off. The first one was named Signal I and is now known to originate from the radical P_{700}^+, the oxidized form of the primary electron donor in Photosystem I. The second signal, "Signal II", however, was 20G broad and remained stable in the dark many hours. It is now known that Commer and co-workers discovered the TyrD$^\bullet$ radical in PSII ('Dark' gave rise to the index "D", which later led to the nomenclature "TyrD"). In the following decade, lasers and improved EPR spectrometers allowed a better kinetic resolution and studies of smaller transient signals which led to the discovery of a new signal

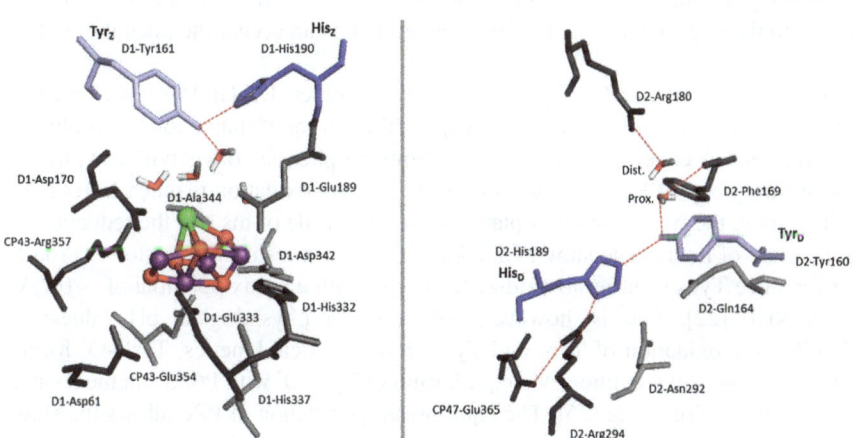

Fig. 4.5 Protein structure around the tyrosines TyrZ and TyrD in Photosystem II. Both tyrosines form hydrogen bonds (red dotted line) to the adjacent histidine residue and to nearby water molecules

with a similar line width (~ 20G): this radical signal decayed in the ms to s time regime after the light was turned off and was named as "Signal II_{fast}". Further work demonstrated that it was present to the same amount as Signal II_D and that the formation and decay kinetics were e.g., pH sensitive. The kinetics of Signal II_{fast} This and its biochemical behaviour allowed the assignment to a kinetic component Z which was discovered earlier by optical spectroscopy (this "Z" was later part of the name "Tyr_Z").

Despite their functional difference, the two tyrosines are both oxidized by $P_{680}{}^{+}$ via proton-coupled electron transfer, resulting in a neutral tyrosine radical $Tyr_{D/Z}^{\bullet}$. Tyr_Z acts as a hole mediator between the Mn_4CaO_5 cluster and the photo-oxidized $P_{680}^{\bullet+}$. Its presence is compulsory for water oxidation, along with the H-bonded partner histidine 190 (His190 or His_Z) [21]. The hydrogen bond length is usually short (2.5 Å), indicating a strong bond. Under physiological conditions, (pH $\approx$ 6.5), it appears that the oxidation of Tyr_Z by $P_{680}^{\bullet+}$ is concerted with deprotonation to His190, resulting in the formation of the pair Tyr_Z-O$^{\bullet}\cdots$ HN^{+}-His190. Upon reduction of Tyr_Z by the Mn-cluster, the transferred proton is transferred back onto Tyr_Z-O$^{\bullet}$ or it exits to the lumen through a hydrogen bond pathway of amino acids and waters [22]. Tyr_Z oxidation is multiphasic, with a fast time component of ~ 10 ns and the longest time component of ~ 0.5 μs. It has been proposed that the fastest component of Tyr_Z oxidation relates to the oxidation of an equilibrated population of Tyr_Z-O$^{-\bullet}\cdots$ HN^{+}-His190 [23]. The slower component has been suggested to involve slower protein motions which promote protein relaxation or proton transfer [24]. The OEC reduces Tyr_Z-O$^{\bullet}$ on the time scale of μs to ms. Due to this fast time scale, it is difficult to study. The redox potential of Tyr_Z has been estimated to be about 1 V versus NHE and the pK_a value about 10.3–12 (Styring et al. 2009). The presence of the OEC seems to expose a strong electrostatic influence to lower the pK_a of His190 by 2–3 log units (4–5) in O_2-evolving Photosystem II. The 'working' pH of PSII is about pH ~ 5.5–7, leading to the suggestions that His190 is neutral and can accept the phenolic proton of Tyr_Z.

D2-Tyr160 (Tyr_D) and its hydrogen bonding partner, D2-His189 (His_D) form a symmetrical counterpart to Tyr_Z and His_D. Although the distance to P_{680} is almost the same (~ 8 Å edge -to - edge distance from the phenolic oxygen of Tyr_D to the nearest ring group of P_{680}, a methyl- group), the Tyr_Z oxidation is much faster than the Tyr_D one: the oxidation takes place on the time scale of ms and the reduction is on the scale of hours. The slow PCET kinetics have to result from a slower proton transfer since Tyr_D is easier to oxidize than Tyr_Z with a redox potential of ~ 0.7 V versus NHE [22]. This is, however, only true for physiological pH values: at pH > 7.7, the oxidation of Tyr_Z and Tyr_D show identical kinetics. TyrD-O$^{\bullet}$ forms under physiological conditions via equilibrium of Tyr_Z-O$^{\bullet}$ with P680$^{\bullet+}$ in the S_2 and S_3 states of the Kok cycle [25]. The equilibrated population of $P_{680}^{\bullet+}$ allows the slow oxidation of Tyr_D-OH which acts as a thermodynamic sink due to its lower redox potential. In contrary to Tyr_Z, which is reduced by the OEC at each step of the Kok cycle, $Tyr_D^{\bullet}$ is reduced by the OEC only in the S_0 state with much slower kinetics. Tyr_D-O$^{\bullet}$ may also be reduced via a slow, long-distance charge recombination process with quinone $Q_A^{\bullet-}$. If the phenolic proton of Tyr_D moves towards His189, it creates a positive charge (H^+N-His189), pushing the location of the hole on $P680^{\bullet+}$ forward onto Tyr_Z, accelerating its oxidation.

4.3 Back-Reactions, Short-Circuits and Leaks: Kinetics and Pathways of Charge Recombination

One of the main challenges encountered by Photosystem II is to find a way to favor energy-conversion processes over competing reactions in which the high energy intermediates decay to the ground state and lower the efficiency of the reaction. This requires the avoidance of back-reactions, short-circuits, by-passes, side-reactions, futile cycles and leaks. PSII mainly deals with this obstacle through kinetic control: forward reactions are faster than backward reactions [26]. Cofactors are appropriately spaced within the protein to secure rapid vectorial electron transfer across the membrane, separating the positive and negative charges from each other. Direct recombination reactions of the radical pairs are strongly exergonic; the standard free energy gap is so big that the reactions are in the 'Marcus inverted region', therefore, they are relatively slow [3]. Additionally, the increased distance between the two charges of the radical pair slows down direct recombination [26].

Furthermore, a simple way to prevent energy loss is to actively slow-down the back-reaction by making it strongly energetically uphill: the shortest route for the electrons to get back to $P^{+\bullet}$ from $Q_B^{\bullet-}$ is e.g., via $Pheo_{D2}$, the pheophytin on the 'non-functional' second branch of the reaction center ("B-side"). The distance is about 9 Å, but the energy gap is thought to be very large, since the $Pheo_{D2}$ midpoint potential is suggested to be more negative than the one of $Pheo_{D1}$ and the potential for the $Q_B/Q_B^{\bullet-}$ couple is about 100 mV more positive than the one of the $Q_A/Q_A^{\bullet-}$ couple [1, 27]. Therefore, a back-reaction via this route is eliminated, contributing to a long life time of the $Q_B^{\bullet-}$ state. A potential back-reaction route for the electron from $Q_B^{\bullet-}$ is via Q_A and since both cofactors are similar in energy, they are already in equilibrium. The next step from $Q_A^{\bullet-}$ to $Pheo_{D1}$ is the step requiring a significant amount of energy: their midpoint potentials differ by several hundred meV [28, 29]. Nevertheless, PSII is able to undergo this route of charge recombination, where the exact pathway is determined by the size of the energy gap between Q_A and $Pheo_{D1}$: charge recombination can be 'direct', from $Q_A^{\bullet-}$ to $P^{+\bullet}$ and 'indirect', via the formation of the $P^{+\bullet}Pheo_{D1}^{\bullet-}$ radical pair. PSII is able to modulate the size of the energy gap and therefore, the yields of these pathways to mitigate protein damage and optimize function [28, 30]. This modulation occurs by switching the redox potential of Q_A: the E_m of $Q_A/Q_A^{\bullet-}$ strongly depends on the presence of the bicarbonate ion (HCO_3^-), a bidentate ligand to the nonheme iron at the electron acceptor site, and the presence of the Mn_4CaO_5 cluster. The absence of the bicarbonate ion results in a shift of the E_m of about 74 mV, from -144 to -70 mV [30].

In the presence of the HCO_3^- ion, the 'indirect' recombination pathway is followed in which the $P^{+\bullet}Pheo^{\bullet-}$ pair decays to the ^3P triplet state [28]. This triplet state lies about 1.3 eV above the ground state and since PSII is far from operating anaerobically, but indeed, produces O_2 itself, the triplet is likely to react with $^3O_2$ to form singlet oxygen ($^1O_2$, 0.98 eV), a highly reactive and damaging species [28]. Most purple bacterial reaction centers are able to quench the ^3P state with carotenoids, which are in van der Waals contact with the bacteriochlorophylls of the reaction center, before it can react with $^3O_2$. In PSII, however, the chlorophylls in the core reaction center are too oxidizing that carotenoids in proximity cannot approach them without being oxidized themselves [31]. In fully-functional PSII, electrons for the $P^{+\bullet}$ reduction are plentifully available due to the water-splitting and the indirect charge recombination route occurs infrequently. Nevertheless, e.g., during photoactivation processes when PSII is already assembled but the Mn_4CaO_5-cluster is still absent, this process is switched off due to the higher redox potential of the $Q_A/Q_A^{\bullet-}$ couple: in the absence of the Mn_4CaO_5 cluster, the structural surrounding of Q_A changes and the $Q_A/Q_A^{\bullet-}$ midpoint potential is about 110 mV more positive than in the functional enzyme [30] (Fig. 4.5).

This results in a larger free energy gap between $P^{+\bullet}Q_A^{\bullet-}$ and $P^{+\bullet}Pheo^{\bullet-}$ and therefore, the other charge recombination route is favored: the 'direct' pathway, where the formation of the troublesome ^3P state is avoided. A similar effect is caused by the absence of the bicarbonate ion: the E_m shift of $Q_A/Q_A^{\bullet-}$ and the consecutive increase of the energy gap between $P^{+\bullet}Q_A^{\bullet-}$ and $P^{+\bullet}Pheo^{\bullet-}$ favors the direct recombination route (Fig. 4.6).

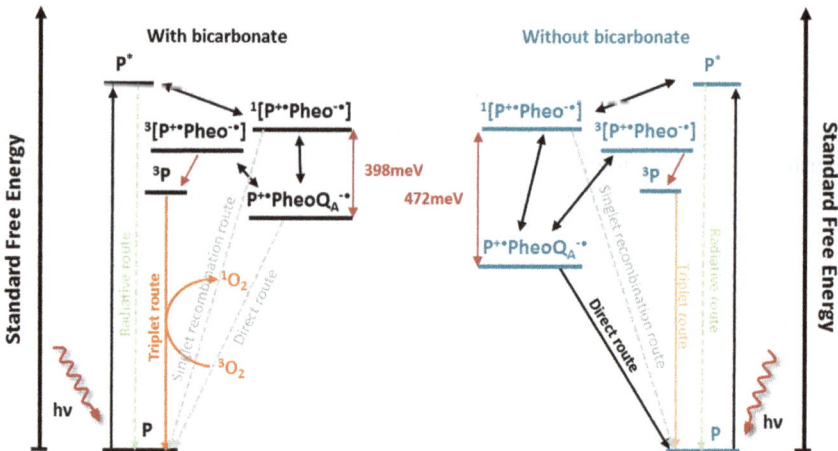

Fig. 4.6 Energy scheme of the charge recombination pathways in PSII and the formation of singlet O_2 influenced by the bicarbonate ion at the electron acceptor site. Two dominant routes can be identified: the triplet (or indirect) route is favoured in the presence of HCO_3^- where the energy gap between $P^{+\bullet}QA^{\bullet-}$ and $P^{+\bullet}Pheo^{\bullet-}$ is small. When HCO_3^- is not bound, the energy gap increases about 74 mV and the direct recombination route is favored. According to Brinkert et al. [30]

Another demonstration of the redox-tuning abilities is shown to be caused by the binding of herbicides to PSII: when phenolic herbicides bind in the Q_B pocket, they not only block the electron transfer from Q_A to Q_B, but they also change the structural surrounding of Q_A by binding via H-bonds to the imidazole which is a ligand to the non-heme iron. This is suggested to weaken the H-bond of the imidazole-Fe-imidazole motif to Q_A on the other side and to cause a change of the Q_A mid-point potential in a way that the 'indirect' recombination route via $P_{D1}^{•+}Pheo_{D1}^{•-}$ is favored [32]. Van Gorkom et al. (1993) suggested that the indirect route is the major recombination pathway at room temperature, which was shown by Rappaport et al. (2002).

Another (indirect) recombination pathway occurs with a yield of approximately 3% [33]: the reformation of P* followed by an exciton decay to the ground state P via emitted fluorescence. All processes are in an equilibrium electron transfer reaction with the S_2 state. Other recombination routes such as the electron transfer from $Pheo_{D1}^{•-}$ to the Mn_4CaO_5 cluster (S_2 state) or $Q_A^{•-}$ to Tyr_Z^{ox} can rather be disregarded due to the distances and involved equilibrium constants.

4.4   Photoelectrochemistry in Semiconductors: Electron Transfer and Recombination Reactions

The charge transfer between a solid (here: metal or semiconductor) in an electrolyte depends on the Franck-Condon principle. It is hereby assumed that the exchange of electrons occurs so quickly that noticeable molecular motions do not happen at the same time: the electrolyte is 'frozen' over this time period ($t < 10^{-14}$ s) [34]. Therefore, electron transfer between electrode and electrolyte molecule occurs before the molecule can react to its new charge: the electron transfer is isoenergetic.

To realize electron transfer from the electrode to the electrolyte, the electrolyte molecule has to possess an unfilled energy level (orbital) at the energetic level of the electron transferred from the electrode. The charger transfer is described as a quantum mechanical tunnelling process between electrode and electrolyte molecules of the outer Helmholtz layer (Fig. 4.7). Therefore, it is also described as an 'outer sphere' charger transfer [34]. Reactions with adsorbed species ('inner sphere reactions') are not accounted for in this model.

In order to determine the current exchange at the electrode, the kinetics of the charge transfer at the solid-electrolyte interface are investigated. Equation (4.5) describes current density (j) as a product of charge (q), reaction rate (k) and concentration (c):

$$j = qkc \tag{4.5}$$

Significant differences between semiconductors and metals result from the gradient of the electric field intensity across the place. Within metals, the potential decrease in the Helmholtz layer results in a potential dependence of the charge

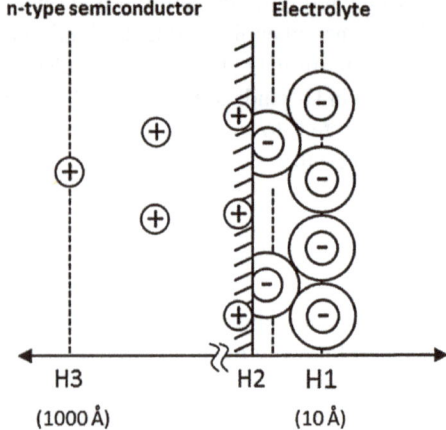

Fig. 4.7 Schematic representation of the charge distribution at the semiconductor electrolyte contact; H1: outer, H2: inner Helmholtz layer, H3: semiconductor edge. According to Lewerenz and Jungblut [34]

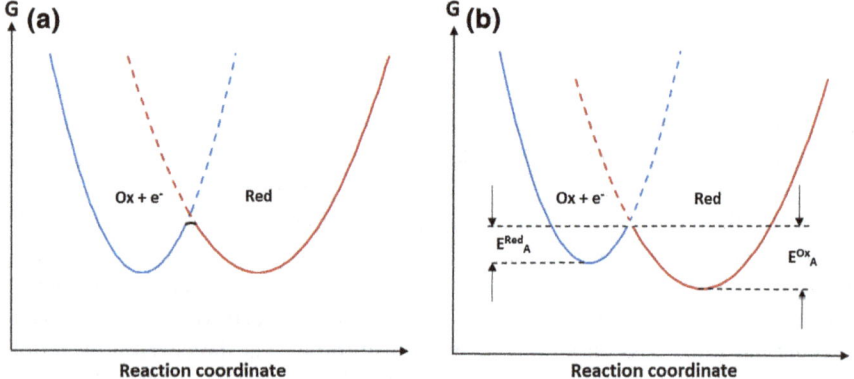

Fig. 4.8 Schematic representation of the energetic relations during charge transfer. According to Lewerenz and Jungblut 34

transfer due to the fact that the electrolyte is energetically shifted with respect to the metal. The relation is generally described in a simplified diagram, in which the Gibb's free energy $G = H-TS$ (H = free enthalpy, T = absolute temperature, S = entropy) is described as a function of the reaction coordinate (Fig. 4.8). Figure 4.8a shows that symmetric processes occur in the case $c_{Ox} = c_{Red}$ (the concentration in the outer Helmholtz layer). In case of a cathodic polarized electrode ($V_e < V_{R,O}$), the electrolyte species is reduced since the activation energies are different for the oxidation and reduction reaction (Fig. 4.8b).

The reaction rates from (4.5) are expressed as:

$$k = k_0 \exp\left(-\frac{E_A}{kT}\right) \tag{4.6}$$

Hereby, k_0 includes the transmission coefficient and the reaction length (scale 5 Å) as well as physical constants. At the metal-electrolyte contact, the activation energy E_A depends on the applied potential since a Galvani potential difference occurs between the metal surface and the outer Helmholtz layer (Fig. 4.9) [34].

Additionally, the location of the activated complexes influences the reaction rate due to the local dependence of the electrical field vertical to the interface. For this reason, the reaction rate of the metal-electrolyte contact depends on the potential according to:

$E_A^{Red}(V) = E_A(0) + \alpha F\Delta V$. Hereby, α is the transfer coefficient with a value from 0 to 1. Experimentally, $\alpha = 0.5$ is observed. Accordingly, the activation energy for the oxidation reaction is expressed as $E_A^{Ox}(V) = E_A(0) - (1-\alpha)F\Delta V$. This results in the potential dependence of the reaction rate (4.6). Equation (4.5) can be written as:

$$
\begin{aligned}
&(1) \quad j_{an}^{(V)} = j_0 e^{-\frac{E_A(0)}{RT}} \cdot e^{-\frac{(1-\alpha)F\Delta V}{RT}} \\
&(2) \quad j_{cath}^{(V)} = j_0 e^{-\frac{E_A(0)}{RT}} \cdot e^{-\frac{\alpha F\Delta V}{RT}}
\end{aligned}
\tag{4.7}
$$

Since $j = j_{an} - j_{cath}$, one obtains the so-called "Butler-Volmer equation", describing the overall current flow at a metal electrode:

$$j(V) = j_0^* \left[e^{\frac{F}{RT}(1-\alpha)\Delta V} - e^{-\frac{F}{RT}\alpha\Delta V} \right] \tag{4.8}$$

ΔV is also often described as the activation overpotential η, since $\Delta V = V_e - V_{eq}$, with V_e being the electrode potential and V_{eq} the equilibrium potential. j_0^* is also known as the exchange current density.

Fig. 4.9 Influence of the electrical field in the Helmholtz layer on the activation energy. (A) Equilibrium situation ($V_e = 0$); (B) electrode potential V_e negative of the equilibrium situation which results in a reduction process. According to Lewerenz and Jungblut 34

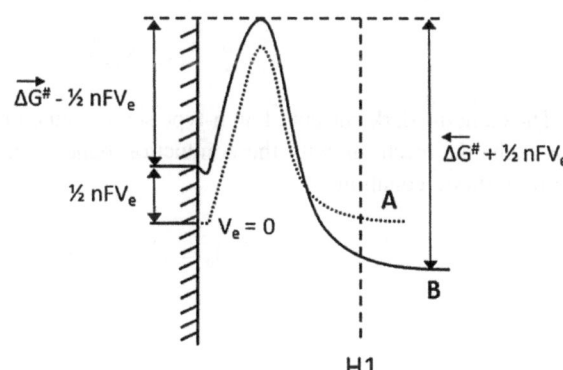

In contrast, the current density in semiconductors is described by the influence of the potential on n_s, the surface concentration of the majority charge carriers. This means that the semiconductor–electrolyte contact behaves in analogy to the semiconductor–metal contact, whereas the respective exchange current densities in equilibrium also include the different physical and chemical processes. The cathodic dark current at an n-type semiconductor can therefore be described as:

$$j(V) = ek_{red}n_s(V) \qquad (4.9)$$

Here, n_s is the electron concentration at the surface $(x = 0)$ and k_{red} is the reaction rate for the reduction. The potential dependence of the current is given by:

$$n_s(V) = n_s(0) \cdot e^{\frac{eV}{kT}} \qquad (4.10)$$

$n_s(0)$ is given by $n_s(0) = N_c e^{\frac{e\varphi_{bh}}{kT}}$, with φ_{bh} being the barrier height of the structure $(E_C - E_F^0)$ and N_c the effective density of states of the lower band edge of the conduction band, given as $N_c = 2\left(\frac{2\pi m_e kT}{h^2}\right)^{3/2}$. This turns (4.9) into:

$$j(V) = ek_{red}N_c e^{\frac{e\varphi_{bh}}{kT}} \cdot e^{\frac{eV}{kT}} \qquad (4.11)$$

Since the barrier height for the flow of electrons from the electrolyte to the semiconductor remains the same, the respective current flow is independent of the applied voltage. For this reason, the current for the case $V = 0$ has to have the opposite sign for the current flow from the semiconductor to the electrolyte:

$$j(0) = -ek_{red}N_c e^{\frac{e\varphi_{bh}}{kT}} \qquad (4.12)$$

The total current density is therefore given as:

$$j(V)_d = j(V) + j(0) = ek_{red}N_c e^{\frac{e\varphi_{bh}}{kT}}\left(e^{\frac{eV}{kT}} - 1\right) \qquad (4.13)$$

The exchange current density in equilibrium is described by the term:

$$j_k^0 = ek_{red}N_c e^{\frac{e\varphi_{bh}}{kT}} \qquad (4.14)$$

The cathodic dark current of an n-type semiconductor and a redox system, which can exchange electrons with the conduction band, is therefore given according to the dark diode equation:

$$j(V)_d = j_k^0\left(e^{\frac{eV}{kT}} - 1\right) \qquad (4.15)$$

The illuminated diode equation is used to describe the voltage generated by the photodiode (the semiconductor), with j_L being the light-induced current (negative (positive) for a photocathode (photoanode)) and n_d being the diode quality factor:

$$j = j_L \pm j_k^0 \left(e^{\pm \frac{eV_{PV}}{n_d kT}} - 1 \right) \tag{4.16}$$

The ideal diode equation assumes that recombination occurs via band to band or recombination via traps in the bulk areas from the device (i.e. not in the junction). Using this assumption, the derivation produces the ideal diode equation above and the quality factor, n_d, is equal to one.

A photoelectrochemical device consists of a current and voltage generating generating photodiode in series with an electrocatalytic overpotential to drive the chemical reaction and a voltage loss term accounting for interfacial, material, solution and/or any other resistances that can described as series elements (Fig. 4.10) [35].

The system voltage is therefore a linear combination of the voltage generated by the photodiode V_{PV} (j), the voltage used by the electrocatalyst η (j) and the one necessary to overcome the system series resistance V_{series}(j):

$$V(j) = V_{PV}(j) + \eta(j) + V_{series}(j) \tag{4.17}$$

Here, V_{PV} can be derived from (4.16). η (J) can be determined by solving the Butler Volmer equation (4.8) and V_{series} (j) can be described as $V_{series} = jR_s$ [35].

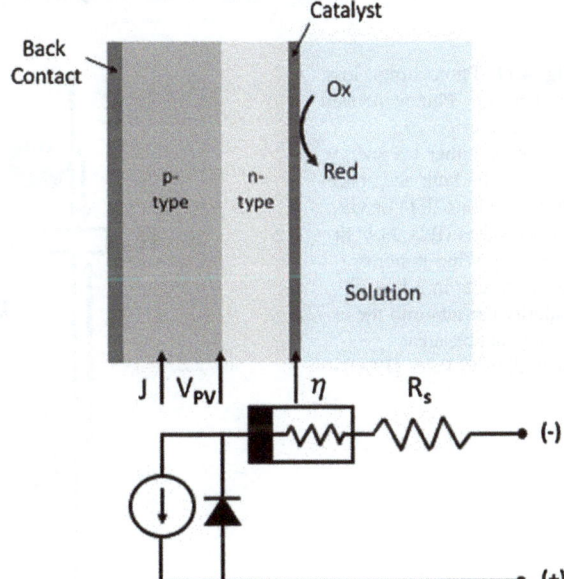

Fig. 4.10 Scheme of a coupled cathodic photodiode-electrocatalyst device. Redrawn from Shaner et al. 2013

Recombination reactions

The quantum efficiency of every photoelectrochemical or photocatalytic process depends on the competition between the desired reaction on one hand and the loss of photogenerated carriers by recombination on the other.

In the case of a light-absorbing semiconductor, electron hole recombination is possible via energy levels in the space charge region (R_1) and at the surface via surface states (R_2, Fig. 4.11). The recombination rate depends on the majority carrier density and therefore, the largest effects are observed at low band-bending, i.e., at potentials close to the flatband, where often no steady-state photocurrent is observed [36].

A number of processes take place at the semiconductor/electrolyte junction on different time scales. Before examining them, two illumination conditions are distinguished: at high light intensities i.e., a high-power laser pulse, more carriers may be generated by the light than were present originally.

This results in a change of the potential drop across the semiconductor due to a redistribution of charge. In an extreme case, the band bending is reduced to zero. At lower light intensities, the perturbation of the charge is much smaller. The collection of photoexcited minority carriers from the space charge and field-free regions of the semiconductor occurs on fast time scales. The transit time for minority carriers in the space charge region is determined by the carrier mobility and the electric field; in the absence of trapping at bulk defect states, the carriers are transferred to the surface in less than a nanosecond.

A minority carrier reaching the interface may be transferred directly to a redox species or trapped by a localized energy level located in the band gap. The rate of these processes can be expressed in terms of the thermal velocity of carriers v, the cross section σ_t and the surface number density N_t of the trapping or redox states:

Fig. 4.11 Photocurrent loss mechanisms. Photogenerated minority carriers can recombine either via energy levels in the bulk and space charge regions (R1) or via surface states (R2). In both cases, capturing minority carrier results in a flux of majority carriers into the recombination area.
According to Peter (1990)

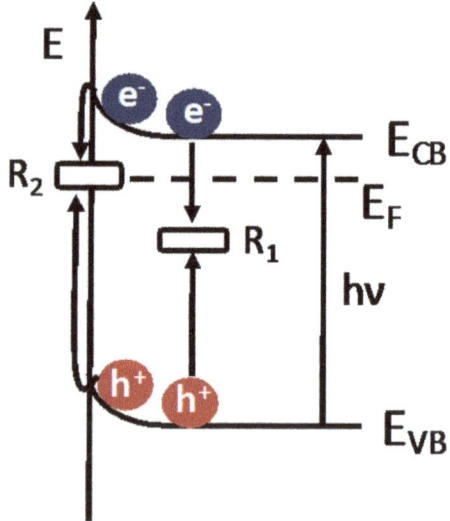

$$s_t = \upsilon \sigma_t N_t \tag{4.18}$$

Here, s_t is described as the surface recombination velocity. It is discussed, however, whether the term should rather be named as minority carrier capture velocity since recombination via surface states is a two-step process which involves also the capturing of majority carriers [36]. Surface recombination velocities can be measured by following the carrier density decay after pulsed laser excitation at open circuit. In the late 1980s, time–resolved photoluminescence measurements investigated e.g., the behaviour of CdS crystals immersed in aqueous electrolyte solutions [37, 38]. The semiconductor/electrolyte interface acts as an additional sink for photogenerated carriers, which adds to the bulk recombination term. The continuity equation [39] determines the time-dependent carrier concentration, $\Delta n(x, t) = \Delta p(x, t)$:

$$\frac{\partial \Delta n(x, t)}{\partial t} = \frac{D^* \partial^2 \Delta n(x, t)}{\partial x^2} - \frac{\Delta n(x, t)}{\tau_b} + G(x, t) \tag{4.19}$$

The transport term takes only diffusion into account; here, it is assumed that the intensity of the light pulse eliminated band bending. The second term describes the bulk recombination which is assumed to be first order. The last term describes the position-dependent generation of carriers which follow the light absorption profile. Benjamin and Huppert (1988) could show that the photoluminescence decay rate for CdS is enhanced by the adsorption of sulphide ions at the CdS/electrolyte interface. When fitting the experimental data to (4.19), they obtained s_t values as high as 10^6 cm s^{-1} [38]. These high values seem to suggest a low quantum efficiency for photoelectrochemical cells based on the n-CdS/S^{2-} system. Evenor et al. (1986) could show, however that high values of s_t are compatible with high photocurrent conversion efficiencies. Even though, most holes are captured by surface states, the two-step oxidation of sulphide at the n-CdS photoelectrode

$$h^+ + S_{ads}^{2-} \rightarrow S_{ads}^- \tag{1}$$

$$2S_{ads}^- + S_{aq}^{2-} \rightarrow S_{aq}^0 + 2S_{ads}^{2-} \tag{2}$$

competes so efficiently with the back-reaction

$$S_{ads}^- + e^- \rightarrow S_{ads}^{2-} \tag{3}$$

that efficient surface-mediated electron transfer occurs. Estimations of the interfacial reaction rate based on Marcus theory suggests that effective mediated electron transfer occurs unless the reorganization energy of the redox reaction is large.

Time-resolved microwave conductivity measurements have also been used to study excess charge carrier kinetics in different systems (e.g. [40]) in e.g., undoped silicon. Here, the lowest surface recombination velocities were found for wafers treated with bichromate solution ($s_t < 10^3$ cm s^{-1}) and by Yablonovitch et al. (1996) for wafers treated with HF solution: s_t (Si(111)) = 0.25 cm s^{-1} [41].

Transferring and capturing carriers at the surface is evidently a fast process. The photocurrent response, however, contains another contribution: when minority carriers accumulate on the surface, the quasi Fermi level shifts away from its equilibrium value and majority carriers start flowing into the surface where they annihilate the trapped minority carriers. This process depends on the concentration of majority carriers at the surface. The first-order rate constant k_{rec} can be expressed in terms of the thermal velocity υ and the capture cross section σ_Γ if their transport through the space-charge region is not rate-determining:

$$k_{rec} = \upsilon \sigma_\Gamma n_{surf} \tag{4.20}$$

Here, n_{surf} is the surface density of majority carriers which can be calculated according to (4.20):

$$n_{surf} = n_{bulk} \exp\left(\frac{-q\Delta\phi_{SC}}{k_B T}\right) \tag{4.21}$$

Here, the term $q\Delta\phi_{SC}$ describes the band bending. The majority carrier concentration at the surface decreases therefore rapidly as the band bending increases, resulting in decreased recombination. If direct charge transfer from the valence band is neglected, the efficiency of the charge transfer processes involving the oxidation of solution species can be formulated in terms of the rates of competing processes involving the surface concentration of trapped holes, p_{surf} [42]:

$$\eta_{trans} = \frac{k_{trans}p_{surf}}{k_{trans}p_{surf} + k_{rec}p_{surf}} = \frac{k_{trans}}{k_{trans} + k_{rec}} \tag{4.22}$$

k_{trans} and k_{rec} are the first order rate constants for charge transfer and recombination, respectively. In general, charge transfer to outer sphere redox systems is sufficiently fast so that recombination can be neglected. If charge transfer, however, is slow due to the involvement of several transfer steps e.g., such as in oxygen evolution, recombination leads to a delayed onset of photocurrent as a function of applied voltage. Alternative to a fast redox system, a majority carrier scavenger can be used which reacts rapidly and irreversibly. For light-driven catalysis using semiconductors as light absorbers, electrocatalysts are employed on the semiconductor surface to enhance majority carrier scavenging (see Chap. 5).

4.4.1 Nanostructured Photoelectrodes

Recently, work on photoelectrochemical water splitting focused also on the utilization of micro- and nanostructured semiconductor electrodes with works describing nanorods [43], nanotubes [44] and nanostructures of the semiconductor. The main advantage which is commonly associated with a structured electrode compared to a planar system is the decoupling of the directions of light absorption and charge-carrier collection [45]. The distance which a minority carrier can diffuse before recombination, is the diffusion length (L_D), which is defined as:

$$L_D = \sqrt{D\tau} \qquad (4.9)$$

Hereby, τ is the minority-carrier lifetime and D is the minority carrier diffusion coefficient. It is related to the minority-carrier mobility, μ (m^2 V^{-1} s^{-1}) by the Einstein relation:

$$D = \frac{\mu k_B T}{q} \qquad (4.10)$$

Considering a traditional planar solar cell, the direction of light absorption is the same as the direction of charge carrier collection [46]. In order to build an efficient cell, the absorber has to be thick enough to absorb all light. At the same time, it has to possess a sufficient electronic quality (i.e. purity and crystallinity) so that the excited minority carriers which are photogenerated deep within the sample are able to diffuse to the surface where they can be collected. This requires that $L_D \geq 1/\alpha$, whereas α is the absorption coefficient of the semiconductor close to the band gap energy. High purity of semiconductors with few defects acting as recombination sites are required to achieve sufficient diffusion lengths in a planar geometry. If nonplanar geometries such as semiconductor rod arrays are used, the required diffusion length can be decoupled from the absorption length (Fig. 4.12b). Device physics modelling supported by experimental results demonstrate that a high surface area semiconductor structure reduces the distance which minority carriers have

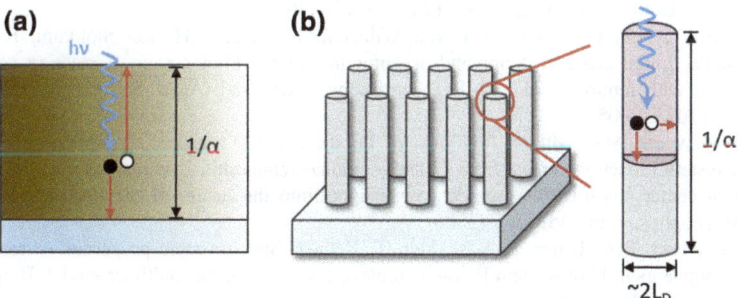

Fig. 4.12 In a planar device such as in (A), photogenerated carriers have to pass through the entire thickness of the cell ($\sim 1/\alpha$) before they are collected (α = absorption coefficient). In a rod-array electrode (B), the carriers only have to reach the rod surface before recombination. Here, LD is the diffusion length of the photogenerated minority carrier (open circle). According to Walter et al. (2010)

to travel and therefore, enable high collection efficiencies despite short minority carrier diffusion lengths.

One major disadvantage of nano- and microstructuring of photoelectrodes is that it reduces the V_{OC} of the dark and light currents [45]. This results from the dependency of the V_{OC} on dark and light current (see Sect. 9.3.2): upon an increased junction area, the V_{OC} decreases due to the reduced splitting of the quasi-Fermi-levels when the photogenerated charge carriers are diluted over a large junction area [47]. Per order of magnitude in increase of the junction area (i.e., solution - semiconductor contact), the photovoltage is predicted to decrease by ≥ 60 mV [46]. In order to achieve a high-performance rod-array electrode, the junction area can be enhanced to collect all carriers with a maximum radius of L_D, but not more in order to account for high V_{OC} losses despite an increased light absorption.

Furthermore, nanostructured semiconductor surfaces reduce electrocatalytic losses in form of overpotentials due to a lower current flux per real area of the electrode. This could be beneficial for earth-abundant catalysts with lower activities which could be spread over a nanostructured electrode.

References

1. F. Rappaport, B.A. Diner, Primary photochemistry and energetics leading to the oxidation of the Mn$_4$Ca-cluster and to the evolution of molecular oxygen in Photosystem II. Coordin. Chem. Rev. **252**(3–4), 259–272 (2008)
2. J.J. Warren, M.E. Ener, A. Vlček Jr., J.R. Winkler, H.B. Gray, Electron hopping through proteins. Coordin. Chem. Rev. **256**, 2478–2487 (2012)
3. R.A. Marcus, On the theory of oxidation-reduction reactions involving electron transfer. I. J. Chem. Phys. **24**(5), 966–978 (1956)
4. R.A. Marcus, N. Sutin, Electron transfer in chemistry and biology. Biochim. Biophys. Acta **811**, 265–322 (1985)
5. V.G. Levich, R.R. Dogonadze, Theory of non-radiative electron transitions in solution from ion to ion. Dokl. Akad. Nauk SSSR **124**, 123–126 (1959)
6. M.L. Groot, N.P. Pawlowicz, L.J. van Wilderen, J. Breton, I.H. van Stokkum, R. van Grondelle, Initial electron donor and acceptor in isolated Photosystem II reaction centers identified with femtosecond mid-IR spectroscopy. Proc. Natl. Acad. Sci. U.S.A. **102**(37), 13087–13092 (2005)
7. A.R. Holzwarth, M.G. Muller, J. Niklas, W. Lubitz, Ultrafast transient absorption studies on Photosystem I reaction centers from Chlamydomonas reinhardtii. 2: mutations near the P700 reaction center chlorophylls provide new insights into the nature of the primary electron donor. Biophys. J. **90**, 552–565 (2006)
8. G. Raszewski, B.A. Diner, E. Schlodder, T. Renger, Spectroscopic properties of reaction center pigments in Photosystem II core complexes: revision of the multimer model. Biophys. J. **95**(1), 105–119 (2008)
9. N. Cox, J.L. Hughes, R. Steffen, P.J. Smith, A.W. Rutherford, R.J. Pace, E. Krausz, Identification of the QY excitation of the primary electron acceptor of Photosystem II: CD determination of its coupling environment. J. Phys. Chem. B **113**(36), 12364–12374 (2009)

10. G. Raszewski, W. Saenger, T. Renger, Theory of optical spectra of Photosystem II reaction centers: location of the triplet state and the identity of the primary electron donor. Biophys. J. **88**(2), 986–998 (2005)
11. R.N. Frese, M. Germano, F.L. de Weerd, I.H. van Stokkum, A.Y. Shkuropatov, V.A. Shuvalov, H.J. van Gorkom, R. van Grondelle, J.P. Dekker, Electric field effects on the chlorophylls, pheophytins, and beta-carotenes in the reaction center of Photosystem II. Biochemistry **42**(30), 9205–9213 (2003)
12. G. Renger, A.R. Holzwarth, Primary electron transfer, in *Photosystem II: The Light-Driven Water Plastoquinone Oxidoreductase*, ed. by T. Wydrzynski, K. Satoh (Springer, Dordrecht, 2005)
13. S. Hammes-Schiffer, A.V. Soudackov, Proton-coupled electron transfer in solution, proteins, and electrochemistry. J. Phys. Chem. B. **112**(45), 14108–14123 (2008)
14. H.B. Gray, J.R. Winkler, Electron tunnelling through proteins. Q. Rev. Biophys. **36**, 341–372 (2003)
15. S.Y. Reece, D.G. Nocera, Proton-coupled electron transfer in biology: results from synergistic studies in natural and model systems. Annu. Rev. Biochem. **78**, 673–699 (2009)
16. R.I. Cukier, Mechanisms for proton-coupled electron-transfer reactions. J. Phys. Chem. **106**, 8442–8454 (1994)
17. S. Hammes-Schiffer, Hydrogen tunnelling and protein motion in enzyme reactions. Acc. Chem. Res. **39**, 93–100 (2006)
18. S. Hammes-Schiffer, Theoretical perspectives on proton-coupled electron transfer reactions. Acc. Chem. Res. **34**, 273–281 (2001)
19. S. Styring, J. Sjöholm, F. Mamedov, Two tyrosines that changed the world: interfacing the oxidizing power of photochemistry to water-splitting in Photosystem II. Biochim. Biophys. Acta **1817**, 76–87 (2012)
20. B. Commer, J. Heise, J. Townsend, Light-induced paramagnetism in chloroplasts. Proc. Natl. Acad. Sci. U.S.A. **43**, 710–718 (1956)
21. T.J. Meyer, M.H.V. Huynh, H.H. Thorp, The possible role of proton-coupled electron transfer (PCET) in water oxidation by Photosystem II. Angew. Chem. Int. Ed. **46**, 5284–5304 (2007)
22. A. Migliore, N.F. Polizzi, M.J. Therien, D.N. Beratan, Biochemistry and theory of proton-coupled electron transfer. Chem. Rev. **114**, 3381–3465 (2014)
23. K. Saito, J.R. Shen, T. Ishida, H. Ishikita, Short hydrogen bond between redox-active tyrosine Y(Z) and D1-His190 in the Photosystem II crystal structure. Biochemistry **50**(45), 9836–9844 (2011)
24. F. Rappaport, A. Boussac, D.A. Force, J. Peloquin, M. Brynda, M. Sugiura, S. Un, R.D. Britt, B.A. Diner, Probing the coupling between proton and electron transfer in Photosystem II core complexes containing a 3-fluorotyrosine. J. Am. Chem. Soc. **131**(12), 4425–4433 (2009)
25. A.W. Rutherford, A. Boussac, P. Faller, The stable tyrosyl radical in Photosystem II: why D? Biochim. Biophys. Acta **1655**(1–3), 222–230 (2004)
26. A.W. Rutherford, A. Osyczka, F. Rappaport, Back-reactions, short-circuits, leaks and other energy wasteful reactions in biological electron transfer: redox tuning to survive life in O_2. FEBS Lett. **586**(5), 603–616 (2012)
27. M.C. Wakeham, M.G. Goodwin, C. McKibbin, M.R. Jones, Photo-accumulation of the $P^+Q_B^-$ radical pair state in purple bacterial reaction centres that lack the Q_A ubiquinone. FEBS Lett. **540**(1–3), 234–240 (2003)
28. G.N. Johnson, A.W. Rutherford, A. Krieger, A change in the midpoint potential of the quinone QA in Photosystem II associated with photoactivation of oxygen evolution. Biochim. Biophys. Acta **1229**(2), 202–207 (1995)
29. F. Rappaport, M. Guergova-Kuras, P.J. Nixon, B.A. Diner, J. Lavergne, Kinetics and pathways of charge recombination in Photosystem II. Biochemistry **41**(26), 8518–8527 (2002)
30. K. Brinkert, S. De Causmaecker, A. Krieger-Liszkey, A. Fantuzzi, A.W. Rutherford, Bicarbonate-induced redox tuning in Photosystem II for regulation and protection. Proc. Natl. Acad. Sci. U.S.A. **113**(43), 12144–12149 (2016)

31. H.J. Van Gorkom, J.P. Schelvis, Kok's oxygen clock: what makes it tick? The structure of P_{680} and consequences of its oxidizing power. Photosynth. Res. **38**(3), 297–301 (1993)

32. R. Takahashi, K. Hasegawa, A. Takano, T. Noguchi, Structures and binding sites of phenolic herbicides in the Q_B pocket of Photosystem II. Biochemistry **49**(26), 5445–5454.De (2010)

33. B.G. Grooth, H.J. Van Gorkom, External electric field effects on prompt and delayed fluorescence in chloroplasts. Biochim. Biophys. Acta **635**, 445–456 (1981)

34. H.-J. Lewerenz, H. Jungblut, *Photovoltaik* (Springer, Berlin, 1995)

35. M.R. Shaner, K.T. Fountaine, H.-J. Lewerenz, Current-voltage characteristics of coupled photodiode-electrocatalyst devices. Appl. Phys. Lett. **103**, 143905–143909 (2013)

36. L.M. Peter, Dynamic aspects of semiconductor photoelectrochemistry. Chem. Rev. **90**, 753–769 (1990)

37. M. Evenor, S. Gottesfeld, Z. Harzian, D. Huppert, S.W. Feldberg, Time-resolved photoluminescence in the picosecond time domain from cadmium sulfide crystals immersed in electrolytes. J. Phys. Chem. **88**, 6213–6218 (1984)

38. D. Benjamin, D. Huppert, Surface recombination velocity measurements of cadmium sulfide single crystals immersed in electrolytes. A picosecond photoluminescence study. J. Phys. Chem. **92**(16), 4676–4679 (1988)

39. J. Vaitkus, The nonequilibrium hall effect and related transport phenomena in semiconductors under inhomogeneous excitation by a laser pulse. Phys. Status Solidi A **34**, 769–775 (1976)

40. M. Kunst, W. Jaegermann, D. Schmeisser, Chemical etching of p-type Si(100) by $K_2Cr_2O_7$. Appl. Phys. A **42**(1), 57–64 (1987)

41. E. Yablonovitch, D.L. Allara, C.C. Chang, T. Gmitter, T.R. Bright, Unusually low surface-recombination velocity on silicon and germanium surfaces. Phys. Rev. Lett. **57**, 249–252 (1986)

42. L.M. Peter, K.G.U. Wijayantha, Photoelectrochemical water splitting at semiconductor electrodes: fundamental problems and new perspectives. Chem. Phys. Chem. **15**, 1983–1995 (2014)

43. T. Lindgren, H. Wang, N. Beermann, L. Vayssieres, A. Hagfeldt, E.-S. Lindquist, Aqueous photoelectrochemistry of hematite nanorod array. Sol. Energy Mater. Sol. Cells **71**, 231–243 (2002)

44. K. Shankar, G.K. Mor, H.E. Prakasam, S. Yoriya, M. Paulose, O.K. Varghese, C.A. Grimes, Highly-ordered TiO_2 nanotube arrays up to 220 μm in length: use in water photoelectrolysis and dye-sensitized solar cells. Nanotechnology **18**(6), 065707 (2007)

45. B.M. Kayes, H.A. Atwater, N.S. Lewis, Comparison of the device physics principles of planar and radial p-n junction nanorod solar cells. J. Appl. Phys. **97**(11) (2005)

46. M.G. Walter, E.L. Warren, J.R. McKone, S.W. Boettcher, Q. Mi, E.A. Santori, N.S. Lewis, Solar water splitting cells. Chem. Rev. **110**, 6446–6473 (2010)

47. M.X. Tan, C.N. Kenyon, N.S. Lewis, Experimental measurement of quasi-fermi levels at an illuminated semiconductor/liquid contact. J. Phys. Chem. **98**, 4959–4962 (1997)

Chapter 5
Water Oxidation Catalysis and Hydrogen Evolution

The application of electrolysis or photoelectrolysis of water to generate oxygen and hydrogen gas could provide a scalable mechanism to store intermittent renewable energy. Hydrogen gas is an energy dense chemical (120 MJ/kg) which can be used in a fuel cell and it is an important feedstock for the chemical industry in processes such as petroleum refining, Fischer-Tropsch synthesis of hydrocarbons and the Haber-Bosch generation of ammonia. Most hydrogen is produced nowadays by steam-reforming of fossil fuels with CO_2 as a byproduct. The production of hydrogen by water electrolysis could in principle be CO_2 emission-free if the electricity was derived from renewables such as solar.

In nature, water oxidation in PSII is required to convert carbon dioxide into organic molecules while at the same time, oxygen is generated to form our aerobic atmosphere. It has long been known that the catalytic water-splitting site contains four Mn ions [1] and understanding their organisation and interaction during the catalytic process has been entitled as the 'holy grail' of bioinorganic chemistry.

This chapter on water oxidation catalysis and hydrogen evolution presents recent insights into the catalytic processes in both, natural and artificial photosynthesis. It elucidates the structure and function of nature's water-splitting catalyst, a heteronuclear Mn_4O_5Ca cluster which catalyzes the reaction in a four-electron oxidation mechanism triggered by photonic absorption. The central functions of cofactors and amino acids in proximity to the cluster are elaborated and recent insights into the catalytic mechanism are compared to the photoelectrocatalytic oxidation of water using photoanodes of transition metal oxides and semiconducting photoanodes with attached electrocatalysts. Brief insights are given as well into recent advances into water oxidation using molecular catalysts based on the transition metals manganese, ruthenium, iridium, iron and cobalt.

Following the introduction of water oxidation mechanisms at the photoanode of a photoelectrochemical cell, the photoelectrochemical generation of hydrogen using semiconducting photocathodes is discussed. Although several semiconductors have band-edge positions which are appropriate for the electrochemical reduction of hydrogen, kinetics for the hydrogen evolution reaction on the bare semiconductor

© Springer International Publishing AG, part of Springer Nature 2018
K. Brinkert, *Energy Conversion in Natural and Artificial Photosynthesis*,
Springer Series in Chemical Physics 117,
https://doi.org/10.1007/978-3-319-77980-5_5

surface generally limit the efficiency of this reaction. Overcoming kinetic limitations requires a stronger driving force, i.e., an overpotential to drive the desired chemical reaction. The addition of surface catalysts can improve the kinetics of the reaction. In this subchapter, materials for the light-induced hydrogen generation are discussed along with mechanisms and theories of the reaction.

5.1 Structure and Function of Nature's Oxygen Evolving Complex

X-Ray diffraction studies of improving resolution have progressively refined the view of the three-dimensional topology and connectivity of the OEC in PSII. Most recently, due to the employment of X-Ray Free Electron Laser (XFEL) pulses, the three-dimensional structure of the inorganic cluster in the S_1 state was obtained at 1.95 Å without radiation damage [2].

The inorganic core of the OEC is described as taking the shape of a 'distorted chair', with the base formed by a heterometallic Mn_3CaO_4 cuboidal unit and the backrest by a Mn-O linkage connected to one of the Mn-ions and one of the oxo bridges of the cubane (Fig. 5.1). Four water-derived ligands are directly ligated to metal ions, two at Mn4 and two at Ca. Carboxylate residues of surrounding amino acids presumably stabilize the cluster in the different redox states. This unit is embedded in the D1 protein and is connected to the CP43 protein by one direct ligand. Only one N-donor, His332 (D1-H332), coordinates the cluster at Mn1, which is also monocoordinated by Glu189 (D1-E189). Important residues include the Tyr_Z-His_Z couple in the second coordination sphere, which is the electron transfer gate to the chlorophylls forming P680.

The OEC is a highly optimized catalyst for water oxidation with turnover frequencies in excess of 100 mol of O_2 (mol OEC)$^{-1}$ in the presence of sufficient quantities of electron acceptors [3]—it evolved over billions of years of evolution as biology's one and only catalyst for water oxidation.

5.1.1 Channel Architecture

Additionally, to fulfilling the requirement of precise spatial organisation of the redox active components involved in excitation energy and electron transfer, the folding and structural arrangement of PSII has to serve the need for the tight control of accessibility and water delivery at the OEC, product release and proton transfer. A precise understanding of these types of regulations are still missing, however, crystallographic models [4, 5], noble gas studies [6], pK_a calculations [7] and molecular dynamics studies taking into account the dynamic structure of the channel architecture [8] have already identified possible channels within PSII which

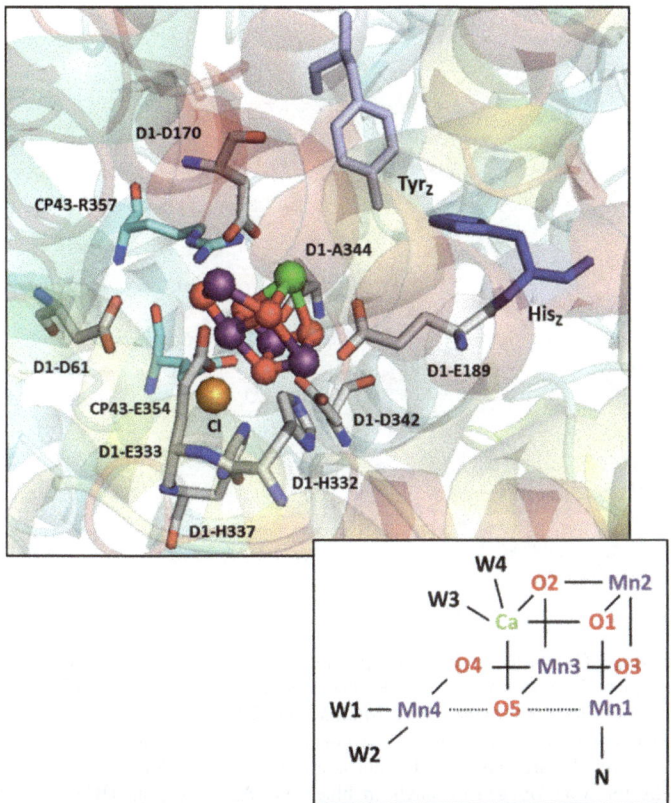

Fig. 5.1 View of the OEC with its immediate environment according to the 1.95 Å structure (PDB reference 4UB6). A common labelling scheme for the inorganic core is indicated in the inset (see text for further explanations)

involve water and O_2 transport and proton transfer (Fig. 5.2). Since the radius of water is 1.49 Å, the assignment is based on the assumption that the radius of a water channel has to be greater than 1.49 ± 0.26 Å [6]. The widest channel has to be the one for oxygen, because the oxygen radius is 1.52 Å and the one of dioxygen is 2.13 Å (along the O=O bond). It has to be the most hydrophobic (since oxygen is hydrophobic) and desirably the shortest one in order to provide fast oxygen removal from the Mn_4CaO_5 cluster. The most significant parameter for the proton channels is the length of the bottleneck (channel narrows) since it should not be longer than the maximal length of a hydrogen bond, 3.5 ± 0.26 Å. In terms of the radius, all channels, which are not wide enough for water or dioxygen transport (radii below 1.49 ± 0.26 Å), but which are still capable of accommodating water molecules, were postulated as possible proton channels [6]. Five solvent-accessible channels were assigned to originate at the OEC, whereas it is not possible to exclusively determine the role of each individual channel. Nevertheless, it appears likely that

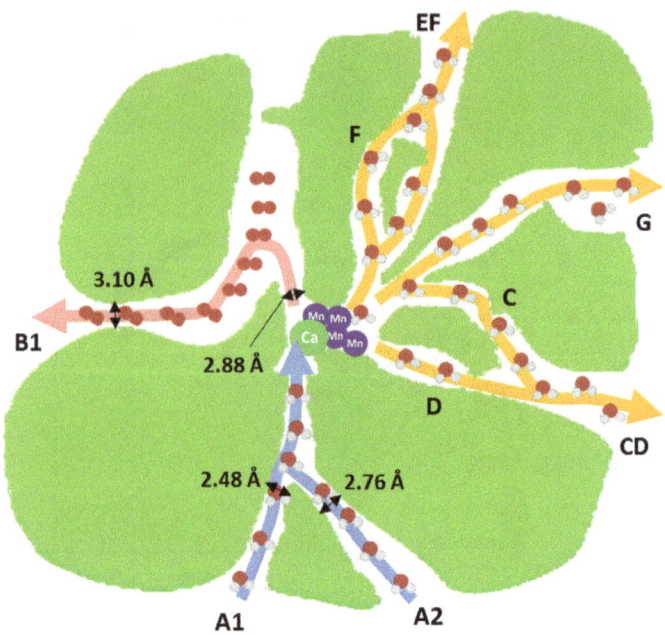

Fig. 5.2 The Mn_4CaO_5 cluster of PSII and possible trajectories of substrate/product channels leading to the lumen. Schematic view from the stromal side onto the membrane plane showing the Mn_4Ca cluster (only Ca (green sphere), Mn1, Mn2, Mn3 and Mn4 (violet spheres) are visible), and putative channels connecting the cluster to the lumenal side. Minimum diameters of the water/oxygen channels (in Å) are indicated by black arrows. Thick coloured arrows indicate the suggested paths for water/oxygen channels in blue (A1, A2), and pink (B1); possible proton channels (C to G) are in yellow According to Gabdulkhakov et al. [6]

water delivery channels are at least one of the channels which is associated with a terminal Mn ion and one channel which is associated with the Ca^{2+} ion; another channel which involves the Ca^{2+} ion may involve an oxygen release pathway (Fig. 5.3).

Notably, water plays at least three distinct roles here [9]: structural water is the surrounding medium on each side of the thylakoid membrane, ordered water chains are involved in proton transfer and the reactant water it is to be oxidized at the active site of the enzyme.

5.1.2 Catalytic Cycle and Manganese Oxidation States

All oxygen in the atmosphere is derived from photosynthetic water-splitting—without it, the biosphere as we know it would not exist. The appearance of the catalytic Mn_4CaO_5 cluster about 3 billion years ago gave rise to the 'big bang of evolution' and it is not surprising that determining the structure of the cluster and

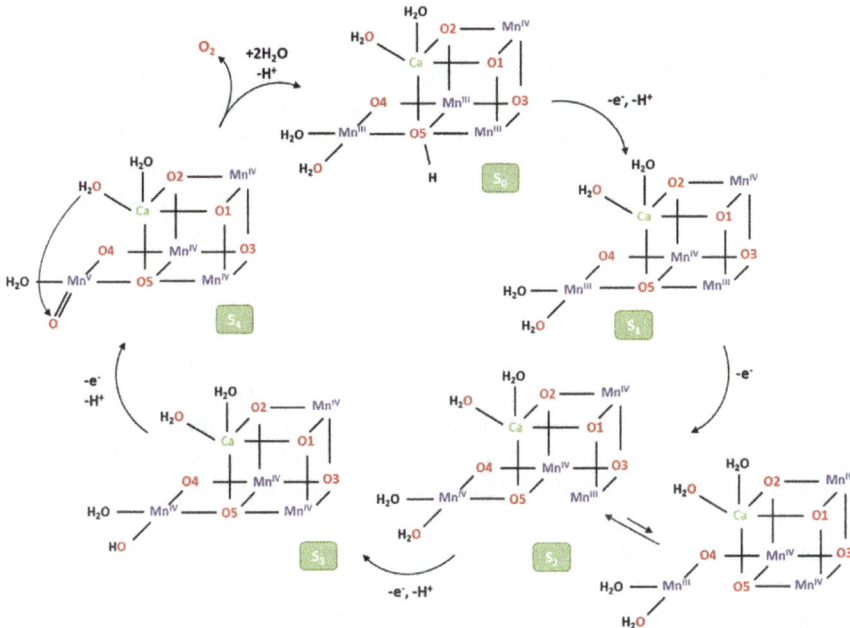

Fig. 5.3 Proposed mechanism by Vinyard et al. [12] of the OEC turnover and substrate water exchange. The S_2 state is described by two interconvertible conformations (see Pantazis et al. [18]). Note: the mechanism for O–O bond formation in the S_4 state is currently under discussion and depends on the structure of the S_4 intermediate

revealing the molecular mechanism of the water-splitting reaction has been one of the greatest challenges in photosynthesis research.

The relative oxidation levels of the OEC are defined by the Kok cycle as described earlier, but it does not include information on the absolute oxidation states of the Mn ions (Sect. 3.1.2.2). Historically, two schemes evolved, accommodating several structural constraints and spectroscopic observations (see e.g. [10, 11]). Since the 1980s it is known that four Mn ions in the oxidation states III and IV compose the active site of the catalyst which give rise to an EPR signal with a spin ground state of $S_{GS} = \frac{1}{2}$ in the S_2 state [10]. This finding could be explained by two oxidation state combinations: either Mn(III)$_3$-Mn(IV) or Mn(III)Mn(IV)$_3$, where in both cases one Mn(III) and one Mn(IV) magnetically couple to yield $S_{dimer} = \frac{1}{2}$ and the remaining two Mn(III) or two Mn(IV) couple to yield $S_{dimer} = 0$, resulting in $S_{GS} = \frac{1}{2}$ in both cases. The two proposals are referred to as the 'low oxidation state' and the 'high oxidation state' schemes or 'low valence' (LV) and 'high valence' (HV) schemes, which differ by two in the total number of electrons in all states. The HV scheme requires the S_0 state to be (III, III, III, IV), whereas the LV scheme corresponds to Mn oxidation states of (II, III, III, III) in the S_0 state up to (III, III, IV, IV) in the S_3 state. Recent computational analyses support the widely accepted hypothesis of the 'high-valence scheme', where the Mn oxidation states are assigned as IV, IV, IV, IV in the S_3 state [9].

At present, the PSII research community agrees in general on the Mn oxidation states of all S states according to the HV scheme [12]. The identification of the Mn oxidation states during the catalytic cycle is central to understanding the principles of biological water oxidation. It places immediate restrictions on the O–O bond formation, since in the HV scheme either oxo–oxyl coupling or nucleophilic attack mechanisms can be imagined, whereas in the LV scheme only the latter is chemically reasonable [9].

Although, a series of X-ray diffraction (XRD) studies have revealed the general structure of the OEC, the atomic structure of any single Kok cycle intermediate could not be determined through these experiments. Early structures obtained by XRD showed reduction of the high-valent Mn centers caused by the X-ray. The availability of metal-metal distances from extended X-ray absorption fine structure (EXAFS) spectroscopy (e.g. [13]) allowed computational studies (QM/MM and density functional theory (DFT), [14, 15]) to refine XRD models to structures corrected for radiation damage.

Generally, there is an agreement on the structures of the S_0, S_1, S_2 and recently also of the S_3 state of the Mn cluster [12, 16]. The decay of the S_4 state is faster than it is formed, which does not allow its investigation as a kinetic intermediate. Therefore, there is no direct experimental evidence for the nature of the O–O bond formation [12]. The inset in Fig. 5.1 shows the resting state of the catalyst (S_1) and is based on a XFEL (X-ray free electron laser) structure. Since the EXAFS results for the S_0, S_1 and S_2 are very similar, it is also thought to be a good model for the S_2 state [17]. Furthermore, the transition from the S_1 to the S_2 state can occur at low temperatures (<200 K) at a fixed protein conformation. It has been shown by theoretical modelling of the S_2 state that the OEC can adopt two interconvertible core topologies in this state, an open cubane and a closed cubane motif. The open cubane structure is slightly lower in energy (ca. 1 kcal/mol) and displays an electronic ground state of spin $S_G = \frac{1}{2}$ at g = 2.0, the so-called "multiline signal". The closed cubane structure, in contrary, expresses an electronic ground state of spin $S_G = 5/2$ at g > 4.1 ("g4 signal"). These two structures show a one-to-one correspondence with two well-known electron paramagnetic resonance (EPR) signals for the S2 state, the multiline and g4 signals [18]. The last metastable intermediate of the reaction cycle is the S_3 state. Recent XFEL and EPR data [18, 19] suggest that all four Mn ions are electronically and structurally similar: they all exhibit a formal oxidation state of +4 and show octahedral local geometry. It has been proposed that an additional water-derived ligand is required at the open coordination site of Mn1, rendering all four Mn ions six-coordinated [17]. The recently obtained XFEL data, however, do not show evidence for an additional water or hydroxo ligand near Mn1 [16], leading to a possible catalytic mechanism as proposed by Vinyard et al. [12] (Fig. 5.3):

5.1.2.1   O–O Bond Formation and Substrate Identification

Although the recent structural results of the Mn cluster described above do not yet constrain the mechanism of water oxidation, they do provide a solid foundation to address the question. In their recent work, Young et al. [16] provided the structure of two-flash illuminated PSII samples (S_3-enriched) and the structure of ammonia-bound two-flash illuminated PSII samples for the investigation of the water-binding site(s). Ammonia was used as a water analogue in several previous studies as it binds to the Mn_4CaO_5 cluster in the S_2 and S_3 states [20].

Due to the remained water-splitting activity in the presence of ammonia, the ammonia-binding Mn site is not a substrate water site, which was used to discriminate between the proposed O–O bond formation mechanisms i.e., whether water oxidation involves an oxo–oxyl radical coupling or a water-nucleophilic attack (Fig. 5.4).

The water-nucleophilic attack mechanism on O5 (Fig. 5.4a, b) requires the involvement of a terminal Ca bound H_2O or OH nucleophile and a formally Mn(V)-oxo or Mn(IV)O˙ (oxyl) electrophile, undergoing an acid-base type reaction (e.g., [21, 22]). This mechanism is also most consistent with the one of synthetic water oxidation catalysts [23] and has been discussed extensively in the past.

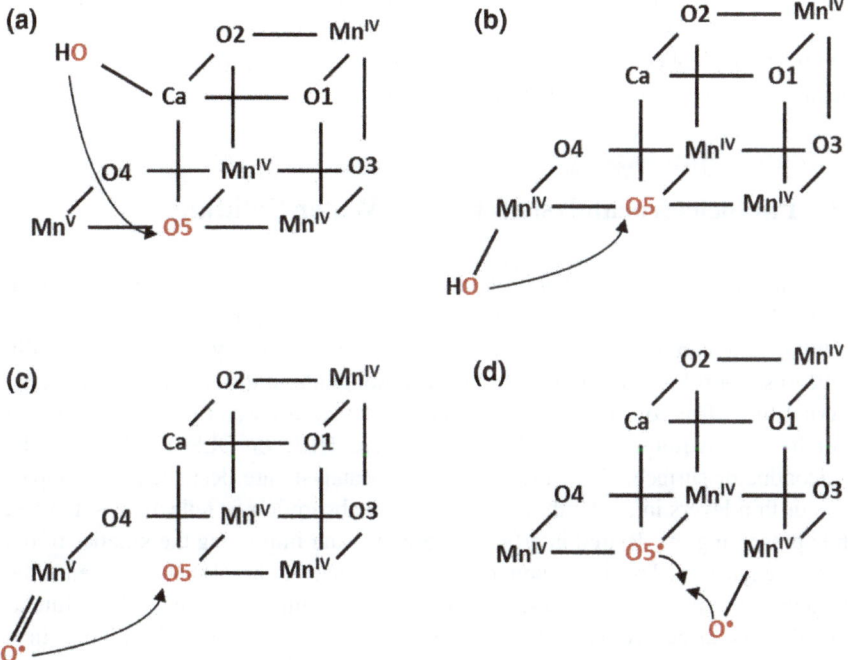

Fig. 5.4 Discussed schemes for the O–O bond formation in PSII in the S_4 state. The models in **a** and **b** involve a nucleophilic attack on O5, whereas the models in **c** and **d** involve radical mechanisms

In the oxo–oxyl radical coupling mechanism (Fig. 5.4c, d), a Mn(IV)-oxyl radical reacts with a Mn-bridging oxo to form O_2. The last scheme (Fig. 5.4d) corresponds to a model by Siegbahn [14, 24], which derived from calculations aiming at yielding the lowest possible energy barriers. Siegbahn suggests that one of the substrate waters first binds during the $S_2 \rightarrow S_3$ transition and is further on oxidized to an oxyl radical in S_4 to carry out the reaction [24]. The position of the involved substrate water position allows to distinguish between the two mechanisms. Recently, Young et al. [16] reported that no evidence was found of the presence of additional water or hydroxo near Mn1 in the XFEL structure of two-flash illuminated PSII samples, which outcompetes direct coupling between a newly bound water-derived ligand in the S3 state on Mn1 and O5 (Fig. 5.4d). This could leave the possibility of O–O bond formation between W3 and O5 (Fig. 5.4a) and W1 and O4 and other relevant mechanisms.

Efforts have been undertaken to determine the kinetics of the substrate water exchange by membrane inlet mass spectrometry (MIMS) measurements followed by rapid mixing with $H_2^{18}O$ [25]. This technique allows the release of $^{34}O_2$ and $^{36}O_2$ in specific S states, probed by single- turnover flashes monitored as a function of incubation time with $H_2^{18}O$. The exchange rate of a terminal water ligand on Mn depends greatly on the protonation, oxidation state, ancillary ligands and geometry.

Unfortunately, until now, there are too little experimental evidences about the S_4 state available which would provide further insights into the key step of the catalytic cycle. New X-ray methods using free electron laser sources allow diffraction patterns to be collected within femtoseconds of X-ray exposure and have the potential to provide more detailed information.

5.2 Photoelectrocatalysis for Solar Water-Splitting

In biomimetic, photoelectrochemical water-splitting cells, many materials which are potentially useful as photocathodes and -anodes for the hydrogen and oxygen evolution reaction do not have surfaces that are sufficiently electrocatalytically active to support light-driven H_2 and O_2 evolution without the application of a large external bias. The construction of an efficient device for water oxidation utilizing sunlight thus requires the attachment of active HER or OER catalysts to the semiconductor surface. Typically, these electrocatalysts are deposited as nanoparticles or thin layers in a way that excessive light absorption or reflection is avoided while preserving the desired interfacial energetics and improving the kinetics of the respective reaction. The attachment of the electrocatalyst particles also influence the energetics of the electron transfer process at the semiconductor surface fundamentally. As discussed in Sect. 3.2.1, excited minority carriers thermalize in a semiconductor liquid junction to the band edge level. When a metallic catalyst is deposited on the semiconductor surface, the minority carriers which participate in the redox reaction with the electrolyte originate from the catalyst particle whose Fermi energy is in equilibrium with the minority carrier quasi-Fermi level of the semiconductor as in a metal-semiconductor Schottky contact [26, 27]. Thus, depositing metal catalysts on the bare semiconductor surface can actually lead to a loss of driving force for the reaction and therefore, it can result in a slower electron transfer rate constant. Nevertheless, this loss in driving force is usually overcome in a higher catalytic turnover rate and furthermore, an increased overall device efficiency.

5.2.1   Electrocatalysts: Oxygen Evolution on Semiconductor Photoelectrodes

In order to carry out water oxidation, electrolysis requires an applied potential >1.23 V between anode and cathode due to the kinetic barriers which are commonly encountered for multistep, multielectron reactions. The definition of this overpotential is simply the voltage applied to the electrode relative to the redox potential of the relevant redox couple in the electrolyte of interest. In case of the photoelectrodes, the electrocatalytic behaviour is intricate with the device properties of the semiconductor/liquid contact and effects the overall performance of a photoelectrolysis cell. A typical, often necessary strategy to improve the performance of photoelectrochemical devices is therefore to add an electrocatalyst to the surface of the semiconductor, deposited as a thin layer or as nanoparticles. A major advance for oxygen evolution catalysts came in 1965, with H. Beer's patent on the dimensionally stable anode (also shortly called DSA, [28]). These electrodes generally consist of RuO_2 and/or IrO_2 and are highly active for electrocatalytic oxidation reactions. These materials are still used today for water oxidation in acidic solutions, however, the anode of choice in commercial electrolyzers remains nickel, operated in hot alkaline solution due to the expense of precious metals [27].

The focus in the development of conductive metal oxides for OER has mainly been on four classes of crystal structures: dioxides, spinels, perovskites and pyrochlores [27]. Typical catalysts include RuO_2 and IrO_2 (first class), Co_3O_4 and $NiCo_2O_4$ (second class), lanthanum oxides such as $NiLa_2O_4$ and $LaCoO_3$ (third class) and $Pb_2Ru(Ir)_2O_7$ (fourth class). Due to their stability, RuO_2 and IrO_2 are mostly used in acid solutions. The question arises, however, when catalysts are attached directly to semiconductor surfaces, which system should be employed. The main differences between catalysts coupled to a light absorbing semiconductors and the ones used in dark electrolysis lies in the fact that the system operating under illumination requires large areas for the maximization in capturing the solar flux. Direct deposition of the catalyst on the light-absorbing semiconductor reduces the requirement for current production per unit geometric area. Commercial electrolysers run therefore at much larger current densities (~ 1 A cm^{-2}) whereas a light-coupled OER system operates in the mA region (e.g., [29]). Of further importance for the choice of the electrocatalyst for HER and OER are stability, cost and earth abundance. The activity of an electrocatalyst is best described and discussed by the corresponding Tafel slope and the $\eta - I$ behaviour, providing a useful tool for evaluating kinetic parameters:

$$\eta = b \log\left(\frac{I}{I_0}\right) \qquad (5.1)$$

Here, η is the overpotential, I is the observed current and I_0 is the exchange current density. The Tafel slope, b, is an indication of the potential increase required to increase the resulting current 1 order of magnitude. Tafel slopes usually reach from 30 to 120 mV/decade whereas the slope of most catalysts increases drastically at higher current densities due to e.g., degradation of the catalyst and uncompensated resistance from bubble formation [27]. The exchange current corresponding to the intercept at $\eta = 0$ is extrapolated from the linear part of the plot of η versus log (I). Qualitatively, I_0 is an indication of how vigorously forward and reverse reactions occur during dynamic equilibrium. On the other hand, b is a measure of how efficiently the electrode responds to an applied potential to produce current. Figure 5.5a compares two hypothetical catalysts with a comparable active surface area (expressed by I_0 resp. J_0) and different electronic activities (expressed by b). Figure 5.5b shows an example of the opposite case.

Ideally, photoelectrocatalysts should not absorb or reflect a significant fraction of incoming light on the surface of a device as discussed above. This is often the case with conductive metal oxide and transition metal catalysts which decrease the overall efficiency. Therefore, the electrocatalyst cannot be deposited as a thick, continuous layer. Either, semiconductor and electrocatalyst have to be micro- or nanostructured to produce a higher surface area for both or transparent catalysts need to be employed (e.g., a transparent conducting oxide). The catalyst deposition represents another obstacle: vacuum (evaporation, sputtering) and solution-phase deposition processes are necessary to protect the light-absorbing semiconductor. Furthermore, direct contact of the semiconductor with the electrocatalyst is required. In the case of semiconductor/liquid junctions it is e.g., important to ensure that highly rectifying or appropriately 'pinched off' contacts are made between

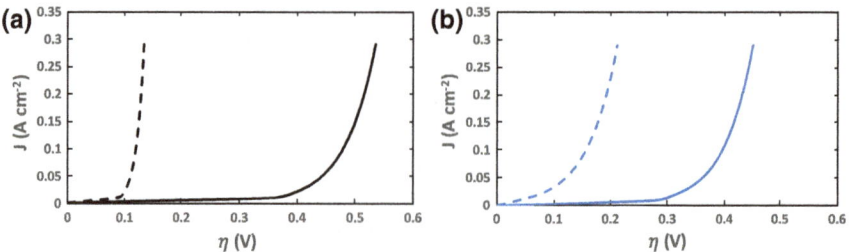

Fig. 5.5 Polarization curve demonstrating the electronic and geometrical effects of a catalyst. **a** Dashed black line: $J_0 = 10^{-5}$ A cm^{-2}, b = 30 mV/decade; solid black line: $J_0 = 10^{-5}$ A cm^{-2}, b = 120 mV/decade. **b** Dashed blue line: $J_0 = 5 \times 10^{-3}$ A cm^{-2}, b = 120 mV/decade; solid blue line: $J_0 = 5 \times 10^{-5}$ A cm^{-2} and b = 120 mV/decade

Fig. 5.6 Volcano plot for
OER catalysts according to
Trasatti [49] in acid. Data
reproduced from Walter et al.
[27]

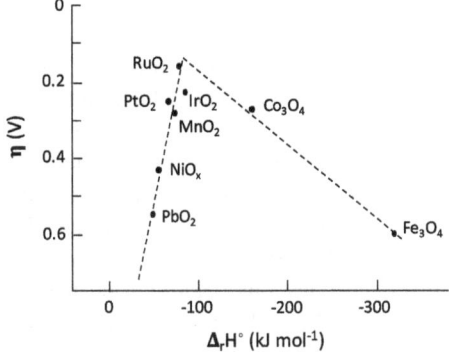

semiconductor and the metal. Beside improving the photoelectrode kinetics, the addition of catalyst particles changes therefore also the energetics of the electron transfer reaction at the semiconductor surface.

In general, the activity of an OER catalyst can be comprehended by the ability of the surface oxide to transition between different valencies to effectively catalyse the oxidation of water. A linear relationship between the minimum potential required for oxygen evolution and their lower oxide/higher oxide redox potentials was discovered by Rasiyah and Tseung, who concluded that O_2 evolution follows a metal oxide redox transition. They formulated the hypothesis that catalysts undergoing such transitions close to the reversible potential for oxygen evolution should possess highest activity [30]. Trasatti [31] related catalytic activity to the metal-oxygen binding strength on the surface of the oxide [31]. The correlation of the overpotential at a fixed current density to the enthalpy of a lower-to-higher-oxide transition results in the so-called "volcano-plot" (Fig. 5.6). Nevertheless, these observations are not related in the literature to any rate-determining step during water oxidation. Rossmeisl et al. [32] derived a rudimentary volcano plot by implementing density functional theory calculations to model the energetics of the OER on rutile-type oxides, relating OER activity to the binding energy of various species on the active oxide surface [32].

5.3   Molecular Catalysts for Water Oxidation

Inspired by the oxygen evolving complex in Photosystem II, many research efforts have been invested in developing a well-defined molecular catalyst for water oxidation which operate in homogenous solution. Such efforts do not only further the understanding of natural water oxidation, but they can also find application in artificial photosynthesis. Numerous OER catalysts for homogenous catalysis have been developed over the past decades (see [3] for an extended review) and a detailed review is beyond the scope of this chapter. Here, the focus should rather lay on trends in the development of molecular OER catalysts.

Probably, the best characterized homogenous catalysts for OER rely on the elements manganese, ruthenium and iridium [33] whereas the first synthetic water oxidation catalyst by Gersten et al. [34] was the so-called 'Blue Dimer', based on ruthenium [34]. Why was ruthenium used at a first instance? Ruthenium complexes are abundantly synthetically accessible and relatively slow ligand exchange rates enable the observation of intermediates. Iridium was known to be the best and stable oxide catalyst for many years, but the synthesis of new compounds was carried out primarily in the last decade. The interest for manganese compounds as water oxidation catalyst does not need much further explanation, following the (still) open questions regarding the manganese coordination chemistry in the catalytic cycle of the $MnCaO_4$ cluster in PSII. Manganese-based model compounds are still of high interest to elucidate the OER in the protein (see Sect. 8.1.3).

Gersten et al. developed the first ruthenium-based water oxidation complex $[Ru^{II}(bpy)_2(py)H_2O]^{2+}$ and emphasised the role of proton-coupled electron transfer (PCET) processes to activate water oxidation [35]. The oxidation of the metal center induces a pK_a shift of water ligands bound to the metal, resulting in the activation of the bound water upon metal oxidation (Fig. 5.7). This results in the formation of hydroxo- and oxo-complexes, advancing the formation of the next oxidation state by promoting the multi-electron oxidation necessary for water oxidation.

The modification of the $Ru(bpy)^{2+}$ series by replacing bpy with 2,2'–bypyrimidine formed a less electron donating analogue. In electrochemical and mechanistic work from 2008, Concepcion et al. reported oxygen evolution occurring from [Ru (terpy)(bpm)H$_2$O]$^{2+}$ (bpm = 2,2'–bipyrimidine, Fig. 5.9a). The mechanism starts with the formation of Ru^{II}-OH$_2$ and the electrochemical oxidation proceeds with a $2H^+/2e^-$ oxidation to form Ru^{IV}=O (Fig. 5.8). Related experiments with cerium(IV)

At pH 0

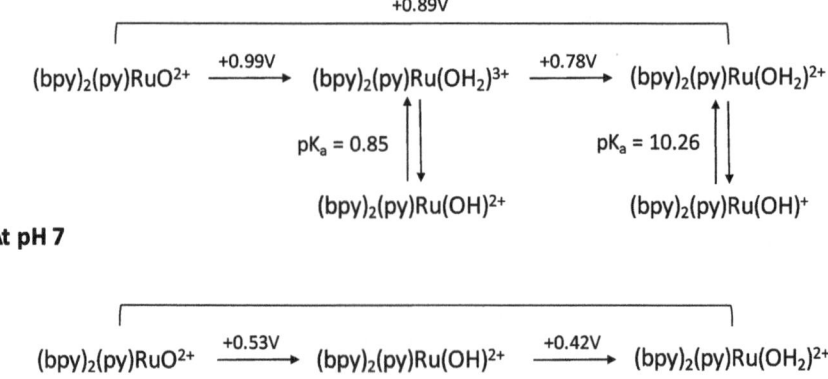

At pH 7

Fig. 5.7 Electrochemical studies by Moyer et al. [35] of $[Ru^{II}(bpy)_2(py)H_2O]^{2+}$. The middle scheme illustrates the processes occurring at pH 0, the bottom scheme shows the processes occurring at pH 7. All redox potentials are reported versus SSCE (SSCE vs NHE +236 mV)

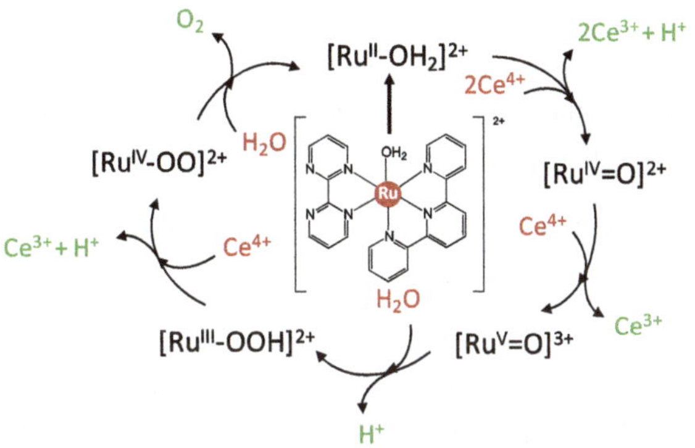

Fig. 5.8 Meyer's mechanism for single-site water oxidation with the $[Ru(terpy)(bpm)H_2O]^{2+}$ complex (according to Concepcion et al. [37]). The rapid oxidation of Ru^{II} forms the key high-valent Ru(V) intermediate. The decay of the peroxidic intermediate Ru^{IV}-OO releasing molecular oxygen is the rate-limiting step

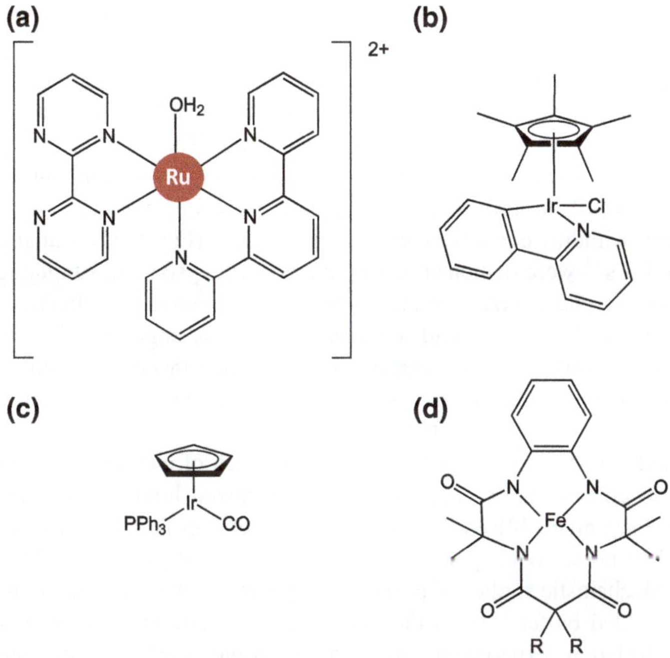

Fig. 5.9 **a** $[Ru(terpy)bpmH_2O]^{2+}$ complex, **b** and **c** [CpIr] complexes and **d** Fe^{III}-TAML complex used for water oxidation. See text for details

show that the formation of the ruthenium(V)-oxo intermediate is able to carry out catalytic water oxidation. In steady-state UV-visible absorption spectroscopy, a side-on peroxide intermediate can be observed. Presumably, this formation is followed by an attack of nucleophilic water on the electrophilic $Ru^V = O$. This species decomposes furthermore to form oxygen and in return $Ru^{II}-OH_2$. The rate-determining step is the decay of the peroxidic intermediate yielding molecular oxygen, whereas the rate of oxygen evolution was determined to be 0.00075 turnovers s^{-1} [36].

The first work on iridium as molecular catalysts for water oxidation was carried out by Bernhard and co-workers. In 2008, the group showed that specific iridium (III) complexes composing of un-substituted or substituted 2-phenylpyridine (ppy) ligands were suitable catalyst precursors [37]. In the first coordination sphere, the compounds consist of two open sites to bind substrate water to the iridium center. Oxygen evolution experiments could confirm a stoichiometric consumption of cerium(IV) coupled to the generation of molecular oxygen. Hull et al. [38] introduced the use of pentamethylcyclopentadienyl (Cp*) as a molecular precatalyst for water oxidation with Ce^{IV} as a sacrificial oxidant (Fig. 5.9b, c [38]). Short times for O_2 evolution rates were found on the order of 10 turnovers min^{-1}, being significantly significantly faster than other systems known at that time.

Iron-based catalysts for water oxidation have been reviewed in detail [39]. The first compounds were developed by Collins in 1980, who worked on the development of robust tetraamido macrocyclic ligands (TAMLs) for iron. They were originally developed for oxidation catalysis with H_2O_2 as the primary oxidant. The key point laid in the employment of a very strong amide ligand, deprotonated by a donor, allowing access to high oxidation state iron. Their application in water oxidation catalysis came in 2010, using CAN (cerium(IV) ammonium nitrate) as the primary oxidant [40]. Improved catalytic performance was achieved with increasingly electron-withdrawing substituents on the TAML (Fig. 5.9d). Initial turnover numbers of 1.3 s^{-1} were demonstrated by the best catalysts with similar activities, whereas a slower phase took over after a few tens of seconds. An intermediacy of the Fe(V) oxy with a 1 e^- oxidized macrocycle was suggested by theoretical analysis of the TAML catalysts with density functional theory and multireference second-order perturbation theory. As the O–O bonding was formed, a water nucleophilic attack was identified [41].

Cobalt and copper based complexes have also been of increasing interest in the search for an efficient OER catalyst. As heterogenized homogenous catalysts, a cobalt phthalocyanine [42] and a fluorinated cobalt corrole [43] have been suggested. Furthermore, cobalt porphyrins have been investigated as OER catalysts (e.g. [44]). Mechanistic studies of porphyrins by Groves et al. showed an important role for the added buffer base in the management of protons to enable catalysis. Catalyst re-isolation experiments and spectroscopic work strongly support a homogenous origin of the catalytic process. This observation is of great interest, since suspensions of simple cobalt salts, phosphate buffer and primary oxidants can already evolve oxygen (e.g. [45]).

Copper complexes have recently attracted attention as catalysts for OER. Mayer and co-workers developed the first Cu-based catalyst with a bipyridine-based system [46]. Despite the relatively large overpotential of ~ 750 mV, required in basic media to drive the catalytic reaction, the catalyst showed a high turnover frequency of ~ 100 s^{-1}. Based on further investigations it has been suggested that the catalyst is a soluble species ligated by bpy and hydroxide. It was estimated that the hydroxyl-substituted bipyridine ligands shows a lower overpotential for the catalytic process [47].

Other interesting, metal-free organic compounds have been investigated for water oxidation. Especially, certain flavin derivatives have been found to be capable of performing oxygen evolution on glassy carbon and platinum electrodes, which was, however, not observed on fluorine-doped tin oxide [48].

Many research efforts have been gathered in the last decades to develop synthetic, inorganic complexes for water oxidation to oxygen. Interestingly, heterogeneous metal oxides based on ruthenium, iridium, iron, cobalt and copper are all known to show activity for water oxidation—whereas nature relies in its water-splitting process on only one element which is able to undergo multiple redox transitions. But the high activity observed in some artificial systems is truly exciting and also encourages further work in improved mechanistic understanding. Natural water-splitting is 2.3 billion years old; but recent progress in research activities provides great hope that we are able to find a competitive system for our energy demands.

5.4 Photoelectrochemical Hydrogen Production

One of the best studied electrochemical reactions is the hydrogen evolution reaction (HER). Similarly to the OER, many materials which are potentially useful as photocathodes do not have sufficiently electrocatalytically active surfaces to support light-assisted hydrogen evolution without the application of a large external bias. Therefore, an efficient water-splitting device requires the attachment of more active HER catalysts to the semiconductor surface. Research efforts have been focused for many years on the development of cheap and efficient water electrolysis systems. Trasatti [52] reviews available heterogeneous catalysts [49]. The HER follows one of two mechanisms [50], each consisting of two primary steps:

$$HA + e^- \rightarrow H^{\bullet *} + A^- \tag{5.2A}$$

$$HA + H^{\bullet *} + e^{-*} \rightarrow H_2 + A^- \tag{5.2B}$$

$$H^{\bullet *} \rightarrow H_2 \tag{5.2C}$$

Step (5.2 A) proceeds in every case, while only one of the other ones (5.2 B or 5.2 C) predominates the further reaction [51]. Here, A$^-$ represents the conjugate base of the reduced (acidic) proton which is H_2O in acidic and OH$^-$ in alkaline media. The asterisk (*) demonstrates a binding site at the electrode surface. Trasatti [52] lists the exchange current densities for HER on pure metals in acid [52]. The exchange current densities plotted versus the metal-hydrogen bond strength gives

Fig. 5.10 Exchange currents for catalytic hydrogen evolution vs strength of metal-hydrogen bond (volcano relation) for pure metals in acidic solution according to Trasatti [52] Redrawn from Walter et al. [27]

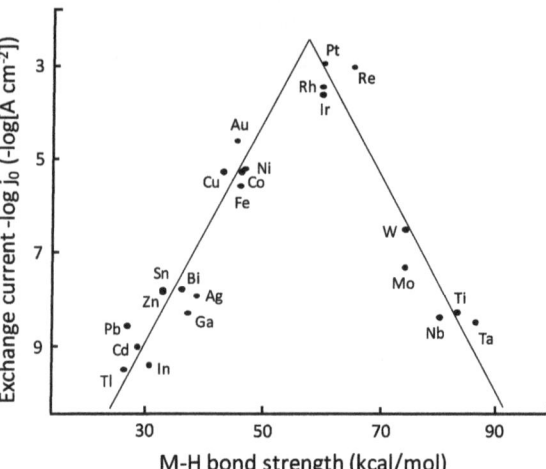

rise to the so-called "volcano plot" (Fig. 5.10), which was already discussed in relation to OER catalysts (see Sect. 5.2.1). The peak value of the HER activity is obtained at intermediate bond strengths and then decreases again at higher bond lengths. Therefore, the most catalytically active precious metal for HER is therefore platinum and the most active ignoble metal is nickel. The fact that based on the volcano relation, the catalytic HER activity arises from the strength of the interaction between the catalyst surface and the adsorbed hydrogen, was independently introduced by Gerischer and Parsons in the 1950s [53, 54]. Parson demonstrated that the three distinct HER steps show a similar pattern i.e. regardless of the predominant mechanisms or the rate-limiting step: the maximum exchange current is always obtained when the free energy of the hydrogen adsorption is zero or close to zero. Further confirmation for the theoretical considerations came from experimental work, showing that protons converted to hydrogen on platinum surfaces bind with small adsorption energies [55]. Recently, Norskøv and co-workers used density functional theory to build a predictive model of HER activities on the basis of calculated adsorption energies [56]. With some accuracy, this model was able to reproduce the volcano curve for metal catalysts, despite the general difficulty to use any fundamental parameter of a material to predict hydrogen adsorption energies and resulting HER activities.

5.4.1   Catalyst Materials and Reaction Mechanisms

Two main compound groups have been studied beside pure metal catalysts for hydrogen evolution: metal composites/alloys and compounds that incorporate nonmetallic elements. Beside the review by Walter et al. [27], an extended review was published in 1992 by Trasatti and only a few highlights should be discussed here. Platinum and other noble metals such as rhodium and ruthenium have been employed and investigated extensively as catalyst materials. Solar-to-hydrogen efficiencies of up to $\sim 13\%$ were achieved in proof-of-concept photoelectrochemical HER systems [57]. Recently, Zong et al. reported on a MoS_2 co-catalyst, producing hydrogen more efficiently than CdS coated with a platinum co-catalyst under illumination and in the presence of a sacrificial reductant [58]. Despite a significant amount of electrolysis literature, only a few samples of working photoelectrochemical HER photocathodes have however employed catalysts other than noble metals. High performance has been demonstrated by metallic nickel after treatments to increase its surface area [59]. Of further particular importance are binary mixtures of Ni-Mo, Ni-Co as well as Ni-Mo-Cd and Ni-Mo-Fe [27]. Geometric, electronic and mixed (synergistic) effects have shown in electrochemical investigations an enhancement of the catalytic activity. The electrochemistry of a set of metal oxides, mostly based on RuO_2, was studied by Kodintsev and Trasatti [60], showing the highest activity [60].

Key parameters in the search for new HER catalysts are beside electrocatalytic activity, stability and long-term performances. Broadly speaking, three parameters influence the photoelectrode and catalyst stability: (i) corrosion, which has a significant impact on the electrode activity over long periods of time; (ii) the poisoning of catalysts by solution impurities and (iii) changes in the electrode composition and/or morphology, occurring on short or long-time scales [27]. The three degradation mechanisms are strongly influences by the storage and operating conditions of the electrode as well as on the electrolyte composition and the pH, cell housing, temperature, potential and current density. Therefore, procedures to minimize degradation require an individual development according to the respective material and its operating conditions.

Presently, there are many HER or OER catalyst options available. Particular concerns are, however, involved with the attachment of these catalysts directly to the semiconductor surface. As illustrated, photoelectrocatalysis required light-absorbers with large areas to maximize the capture of solar flux. If the catalyst is deposited directly on the semiconductor surface, the requirements for current production per unit geometric area is drastically reduced. A larger electrode area also requires a larger amount of catalyst in order to cover the area, bearing the question of materials costs. Furthermore, the catalyst should not absorb light itself or reflecting the incoming sunlight: a thick continuous layer of catalyst with a high surface area as found in industrial electrolyzers can therefore not be employed in semiconductor-coupled systems since the metallic overlayer would absorb or reflect most of the incoming light. The development of transparent catalysts such as a transparent conductive oxide or the employment of a system in which both, the absorber and the catalyst are micro- or nanostructured to increase the surface area for both parts are solutions which are currently investigated (see Sect 4.4.1, [27]).

References

1. R.J. Debus, The manganese and calcium ions of photosynthetic oxygen evolution. Biochim Biophys. Acta. Bioenerg. **1102**, 269–352 (1992)
2. M. Suga et al., Native structure of Photosystem II at 1.95 Å resolution viewed by femtosecond X-ray pulses. Nature **517**(7532), 99–103 (2015)
3. J.D. Blakemore, R.H. Crabtree, G.W. Brudvig, Molecular catalysts for water oxidation. Chem. Rev. **115**(23), 12974–13005 (2015)
4. J.W. Murray, J. Barber, Structural characteristics of channels and pathways in Photosystem II including the identification of an oxygen channel. J. Struct. Biol. **159**, 228–237 (2007)
5. F.M. Ho, S. Styring, Access channels and methanol binding site to the CaMn₄ cluster in Photosystem II based on solvent accessibility simulations, with implications for substrate water access. Biochim. Biophys. Acta **1777**, 140–153 (2008)
6. A. Gabdulkhakov, A. Guskov, M. Broser, J. Kern, F. Müh, W. Saenger, A. Zouni, Probing the accessibility of the Mn₄Ca cluster in Photosystem II: channels calculation, noble gas derivatization, and cocrystallization with DMSO. Structure **17**(9), 1223–1234 (2009)
7. H. Ishikita, W. Saenger, B. Loll, J. Biesiadka, E.W. Knapp, Energetics of a possible proton exit pathway for water oxidation in Photosystem II. Biochemistry **45**, 2063–2071 (2006)
8. S. Vassiliev, P. Comte, A. Mahboob, D. Bruce, Tracking the flow of water through Photosystem II using molecular dynamics and streamline tracing. Biochemistry **49**, 1873–1881 (2010)
9. V. Krewald, M. Retegan, N. Cox, J. Messinger, W. Lubitz, S. DeBeer, F. Neese, D.A. Pantazis, Metal oxidation states in biological water splitting. Chem. Sci. **6**, 1676–1695 (2015)
10. G.C. Dismukes, Y. Siderer, Intermediates of a polynuclear manganese center involved in photosynthetic oxidation of water. Proc. Natl. Acad. Sci. U.S.A. **93**, 3335–3340 (1981)
11. H. Dau, I. Zaharieva, M. Haumann, Recent developments in research on water oxidation by Photosystem II. Curr. Opin. Chem. Biol. **16**(1–2), 3–10 (2012)
12. D.J. Vinyard, S. Khan, G.W. Brudvig, Photosynthetic water oxidation: binding and activation of substrate waters for O–O bond formation. Faraday Discuss. **185**, 37–50 (2015)
13. K. Sauer, J. Yano, V.K. Yachandra, X-ray spectroscopy of the photosynthetic oxygen-evolving complex. Coord. Chem. Rev. **252**, 318–335 (2008)
14. P.E.M. Siegbahn, A structure-consistent mechanism for dioxygen formation in Photosystem II. Chem. Eur. J. **14**, 8290–8302 (2008)
15. J. Yano, V. Yachandra, Mn₄Ca cluster in photosynthesis: where and how water is oxidized to dioxygen. Chem. Rev. **114**, 4175–4205 (2014)
16. I.D. Young et al., Structure of Photosystem II and substrate binding at room temperature. Nature **540**, 453–457 (2016)
17. M. Pèrez-Navarro, F. Neese, W. Lubitz, D.A. Pantazis, N. Cox, Recent developments in biological water oxidation. Curr. Opin. Chem. Biol. **31**, 113–119 (2016)
18. D.A. Pantazis, W. Ames, N. Cox, W. Lubitz, F. Neese, Two interconvertible structures that explain the spectroscopic properties of the oxygen-evolving complex of Photosystem II in the S2 state. Angew. Chem. Int. Ed. **51**, 9935–9940 (2012)
19. N. Cox, M. Retegan, F. Neese, D.A. Pantazis, A. Boussac, W. Lubitz, Electronic structure of the oxygen-evolving complex in Photosystem II prior to O–O bond formation. Science **345**, 804–808 (2016)
20. A. Boussac, A.W. Rutherford, S. Styring, Interaction of ammonia with the water-splitting enzyme of Photosystem II. Biochemistry **29**, 24–32 (1990)
21. J. Messinger, M. Badger, T. Wydrzynski, Detection of one slowly exchanging substrate water molecule in the S3 state of Photosystem II. Proc. Natl. Acad. Sci. U.S.A. **11**, 3209–3213 (1995)
22. V.L. Pecoraro, M.J. Baldwin, M.T. Caudle, W.-Y. Hsieh, N.A. Law, A proposal for water oxidation in Photosystem II. Pure Appl. Chem. **70**, 925–929 (1998)

23. S. Romain, L. Vigara, A. Llobet, Oxygen-oxygen bond formation pathways promoted by ruthenium complexes. Acc. Chem. Res. **42**, 1944–1953 (2009)

24. P.E.M. Siegbahn, Water oxidation mechanism in Photosystem II, including oxidations, proton release pathways, O–O bond formation and O_2 release. Biochim. Biophys. Acta **1827**, 1003–1019 (2013)

25. W. Hillier, T. Wydrzynski, ^{18}O-Water exchange in Photosystem II: substrate binding and intermediates of the water splitting cycle. Coord. Chem. Rev. **252**, 306–317 (2008)

26. M.X. Tan, C.N. Kenyon, N.S. Lewis, Experimental measurement of quasi-fermi levels at an illuminated semiconductor/liquid contact. J. Phys. Chem. **98**, 4959–4962 (1997)

27. M.G. Walter, E.L. Warren, J.R. McKone, S.W. Boettcher, Q. Mi, E.A. Santori, N.S. Lewis, Solar water splitting cells. Chem. Rev. **110**, 6446–6473 (2010)

28. H.B. Beer, Patent (England), vol. 1, 147–442 (1965)

29. M.M. May, H.-J. Lewerenz, D. Lackner, F. Dimroth, T. Hannappel, Efficient direct solar-to-hydrogen conversion by in situ interface transformation of a tandem structure. Nat. Comm. **6**, 8272–8286 (2015)

30. P. Rasiyah, A.C.C. Tseung, The role of the lower metal oxide/higher metal oxide couple in oxygen evolution reactions. J. Electrochem. Soc. **131**, 803–808 (1984)

31. S. Trasatti, Electrocatalysis by oxides—attempt at a unifying approach. J. Electroanal. Chem. **111**, 125–131 (1980)

32. J. Rossmeisl, Z.W. Qu, H. Zhu, G.J. Kroes, J.K. Nørskov, Electrolysis of water on oxide surfaces. Electroanalyt. Chem. **607**, 83–89 (2007)

33. C.W. Cady, R.H. Crabtree, G.W. Brudvig, Functional models for the oxygen-evolving complex of Photosystem II. Coord. Chem. Rev. **252**, 444–455 (2008)

34. S.W. Gersten, G.J. Samuels, T.J. Meyer, Catalytic oxidation of water by an oxo-bridged ruthenium dimer. J. Am. Chem. Soc. **104**, 4029–4030 (1982)

35. B.A. Moyer, T.J. Meyer, Oxobis (2,2'-bipyridine)-pyridineruthenium (IV) ion, [(bpy)$_2$(py)Ru=O]$^{2+}$. J. Am. Chem. Soc. **100**, 3601–3603 (1978)

36. N.D. McDaniel, F.J. Coughlin, L.L. Tinker, S. Bernhard, Cyclometalated iridium(III) aquo complexes: efficient and tunable catalysts for the homogenous oxidation of water. J. Am. Chem. Soc. **130**, 8730–8731 (2008)

37. J.J. Concepcion, J.W. Jurss, J.L. Templeton, T.J. Meyer, One site is enough. Catalytic water oxidation by [Ru(tpy)(bpm)(OH$_2$)]$^{2+}$ and [Ru(tpy)(bpz)(OH$_2$)]$^{2+}$. J. Am. Chem. Soc. **130**(49), 16462–16463 (2008)

38. J.F. Hull, D. Balcells, J.D. Blakemore, C.D. Incarvito, O. Eisenstein, G.W. Brudvig, R.H. Crabtree, Highly active and robust Cp* iridium complexes for catalytic water oxidation. J. Am. Chem. Soc. **131**, 8730–8731 (2009)

39. A. Singh, L. Spiccia, Water oxidation catalysts based on abundant 1st row transition metals. Coord. Chem. Rev. **257**, 2607–2622 (2013)

40. W.C. Ellis, N.D. McDaniel, S. Bernhard, T.J. Collins, Fast water oxidation using iron. J. Am. Chem. Soc. **132**, 10990–10991 (2010)

41. M.Z. Ertem, L. Gagliardi, C.J. Cramer, Quantum chemical characterization of the mechanism of an iron-based water oxidation catalyst. Chem. Sci. **3**, 1293–1299 (2012)

42. T. Abe, K. Nagai, S. Kabutomori, M. Kaneko, A. Tajiri, T. Norimatsu, Photoelectrode working in the water phase: visible-light-induced dioxygen evolution by a perylene derivative/cobalt phthalocyanine bilayer. Angew. Chem. Int. Ed. Engl. **45**, 2778–2781 (2006)

43. D.K. Dogutan, R. McGuire, D.G. Nocera, Electrocatalytic water oxidation by cobalt(III) hangman β-octafluoro corroles. J. Am. Chem. Soc. **133**, 9178–9180 (2011)

44. T. Nakazono, A.R. Parent, K. Sakai, Porphyrins as homogenous catalysts for water oxidation. Chem. Commun. **49**, 6325–6327 (2013)

45. V.Y. Shafirovich, N.K. Khannanov, V.V. Strelets, Chemical and light-induced catalytic water oxidation. Nouv. J. Chim. **4**, 81–84 (1980)

46. S.M. Barnett, K.I. Goldberg, J.M. Mayer, Soluble copper-bipyride water-oxidation electro-catalyst. Nat. Chem. **4**, 498–502 (2012)

47. T. Zhang, C. Wang, S. Liu, J.-L. Wang, W. Lin, A biomimetic copper water oxidation catalyst with low overpotential. J. Am. Chem. Soc. **136**, 273–281 (2013)
48. E. Mirzakulova, R. Khatmullin, J. Walpita, T. Corrigam, N.M. Vargas-Barbosa, S. Vyas, S. Oottikkal, S.F. Manzer, C.M. Hadad, K.D. Glusac, Electrode-assisted catalytic water oxidation by flavin derivative. Nat. Chem. **4**, 794–801 (2012)
49. S. Trasatti, Electrocatalysis by oxides—attempt at a unifying approach. J. Electroanal. Chem. **111**, 125–131 (1980)
50. B.E. Conway, M. Salomon, Electrochemical reaction orders: applications to the hydrogen-and oxygen-evolution reactions. Electrochim. Acta **9**(12), 1599–1615 (1964)
51. A.J. Nozik, p-n photoelectrolysis cells. Appl. Phys. Lett. **29**(3), 150–153 (1976)
52. S.J. Trasatti, Work function, electronegativity, and electrochemical behaviour of metals: III. Electrolytic hydrogen evolution in acid solutions. Electroanal. Chem. **39**(1), 163–184 (1972)
53. H. Gerischer, Mechanism of electrolytic discharge of hydrogen and adsorption energy of atomic hydrogen. Bull. Soc. Chim. Belg. **67**, 506–512 (1958)
54. R. Parsons, The rate of electrolytic hydrogen evolution and the heat of adsorption of hydrogen. Trans. Faraday Soc. **54**, 1053–1063 (1958)
55. B.E. Conway, B.V. Tilak, Interfacial processes involving electrocatalytic evolution and oxidation of H_2, and the role of chemisorbed H. Electrochim. Acta **47**, 3571–3594 (2002)
56. B. Hinnemann, P.G. Moses, J. Bonde, K.P. Jorgensen, J.H. Nielsen, S. Horch, I. Chorkendorff, J.K. Norskov, Biomimetic hydrogen evolution: MoS_2 nanoparticles as catalyst for hydrogen evolution. J. Am. Chem. Soc. **127**(15), 5308–5309 (2005)
57. E. Aharon-Shalom, A. Heller, Efficient p-InP (Rh-H alloy) and p-InP (Re H alloy) hydrogen evolving photocathodes. J. Electrochem. Soc. **129**(12), 2865–2866 (1982)
58. X. Zong, H. Yan, G. Wu, G. Ma, F. Wen, L. Wang, C. Li, Enhancement of photocatalytic H2 evolution on CdS by Loading MoS_2 as cocatalyst under visible light irradiation. J. Am. Chem. Soc. **130**(23), 7176–7177 (2008)
59. H. Wendt, V. Plzak, Electrocatalytic and thermal activation of anodic oxygen- and cathodic hydrogen-evolution in alkaline water electrolysis. Electrochim. Acta **28**, 27–34 (1983)
60. I.M. Kodintsev, S. Trasatti, Electrocatalysis of H_2 evolution on RuO_2 + IrO_2 mixed oxide electrodes. Electrochim. Acta **39**, 1804–1808 (1994)

Chapter 6
Carbon Fixation

The International Energy Agency predicts a global annual energy consumption of about 28 TW in 2050, whereas the global population is predicted to increase from 7.2 to 9.7 billion. This doubled energy consumption let us face a new challenge in the 21st century: new energy sources are required, which allow us to step away from the use of fossil fuels. Although, energy produced from fossil fuels has great importance due to burning (oxidation to carbon dioxide and water) producing significant amounts of energy per unit weight, it encounters for two major impediments: (i) the alarming consumption of energy assets and (ii) global warming due to increasing atmospheric CO_2 concentrations. Every year, more than 10 billion tons of carbon dioxide is added to our atmosphere [1]. The U.S. Department of Energy estimates that natural processes can only absorb about half of the amount.

Natural photosynthesis uses carbon dioxide as an energy source, incorporating it into long chain hydrocarbons. Recent approaches in artificial photosynthesis research have focused on a similar attempt, using abundant CO_2 for the production of 'solar fuels' by converting it photoelectrocatalytically in a semiconductor-electrocatalyst system into e.g., ethanol. An efficient CO_2 recycling process along this approach could regulate global warming and provide a sustainable fuel supply [2]. Other conversion processes besides the photoelectrocatalytic conversion of CO_2 exist, but include several disadvantages: (i) high temperature and high electrical voltage are required to break down CO_2 molecules, (ii) raw materials are limited, (iii) operation costs are high and (iv) unsustainability.

In the following, the CO_2 fixation process in natural photosynthesis should be elucidated and contrasted to strategies for photoelectrochemical CO_2 reduction. Recent developments in catalysis and insights into mechanistic studies are further on discussed.

© Springer International Publishing AG, part of Springer Nature 2018
K. Brinkert, *Energy Conversion in Natural and Artificial Photosynthesis*,
Springer Series in Chemical Physics 117,
https://doi.org/10.1007/978-3-319-77980-5_6

6.1 RubisCO—Structure, Functionality and Catalytic Efficiencies

The net CO_2 assimilation process in photosynthesis is carried out by the enzyme Ribulose-1,5-bisphosphate (RuBP) carboxylase/oxygenase (RubisCO) as already briefly described in Sect. 2.2. Carbon fixation resulting from RubisCOs activity yields in more than 10^{11} tons of atmospheric CO_2 annually [3]. Nevertheless, at the top of the canopy in field-grown crops, this process is often the rate-limiting step of photosynthesis [4]: although, RubisCO is the most abundant protein on earth [5, 6], it has severe limitations. The enzyme is extremely inefficient and its carboxylation activity is affected by numerous side-reactions, especially with O_2: due to the high $O_2 : CO_2$ ratio in ambient air (approx. 500 : 1), an average RubisCO enzyme fixes only up to two O_2 molecules every five CO_2-fixation reactions [7]. RubisCO depends on effector molecules to modulate its activity and on ancillary proteins such as the RubisCO activase to control its activation state [8]. The correct folding and assembly in the cell is mediated by Chaperones, whereas the exact details of the process vary between enzymes of different origin. Right now, high resolution three-dimensional structures of RubisCO are available from different organisms (reviewed e.g., in [9]). All RubisCO enzymes are multimeric with two different subunits: a large catalytic subunit (L, 50–55 kDa) and a small (S, 12–18 kDa) one. Different molecular forms of RubisCO can be distinguished according to the presence or absence of the small subunit. The most common form (also referred to

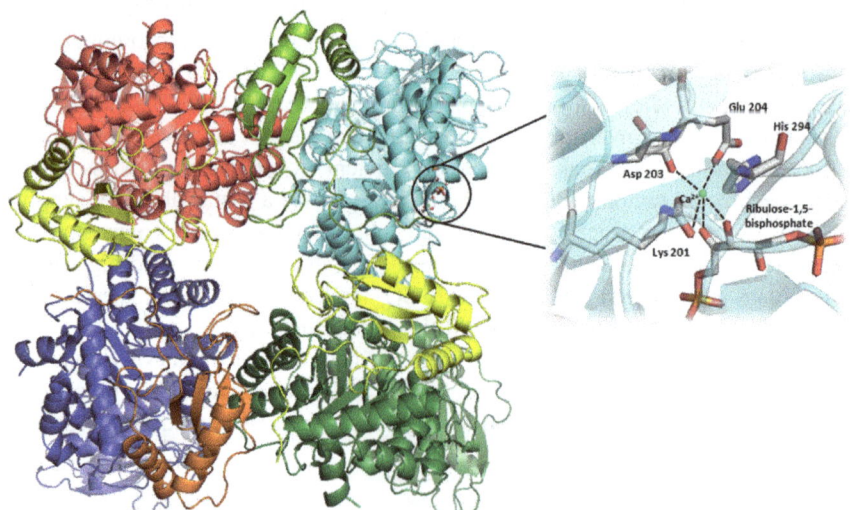

Fig. 6.1 Structure of the activated spinach RuBisCO complex with its substrate ribulose-1,5-bisphosphate and calcium at a resolution of 2.1 Å (pdb reference 1RXO). Amino acid abbreviations are in the standard three-letter code in the inset. The use of calcium instead of magnesium as the activator metal enabled the trapping of the substrate in a stable complex

as 'form I') is composed of large and small subunits in a hexadecameric structure, L8S8 (Fig. 6.1), and is present in most chemoautrophic bacteria, cyanobacteria, red and brown algae and in all higher plants. The core consists of four L2 dimers arranged around a 4-fold axis, covered at each end by four small subunits [10]. The small subunit is not essential for catalysis due to the fact that the large subunit octamer remains carboxylase activity. The form II enzyme lacks small subunits and is a dimer of large subunits $(L2)_n$. Initially, it was discovered in purple, non-sulphur bacteria and several chemoautrophic bacteria. The secondary structure of the large, catalytic unit is extremely well conserved throughout different forms of the enzyme, despite they differ in their amino acid sequence and function [9].

The large subunit of RubisCO is encoded by a single gene in the chloroplast genome and its synthesis is carried out by the plastid ribosome. In plants, the small subunit is coded by a family of nuclear genes and its synthesis occurs in the cytosol [8]. The synthesis and assembly of the holoenzyme including the coordinated control of chloroplastic and cytosolic processes have been demonstrated to require the assistance of ancillary proteins (chaperones).

The pathway involving the reaction of CO_2 and H_2O with RuBP yielding in two molecules of 3PGA can be described by multiple discrete steps and involves associated intermediates of variable stability (Fig. 6.2). For functionality, RubisCO requires activation by carbamylation of the ε-amino group of the active-site Lys201 via a CO_2 molecule. The carbamylated Lys201 is further on stabilized by the binding of a magnesium ion to the carbamate (1). The Mg^{2+} ion interacts in the following with the RuBP, yielding in the formation of a spontaneous complex sequestering Mg^{2+} ((2), [11]). Computations suggest that a concentration level twice as high of RuBP than RubisCO sites leads to a 90% activation level of the enzyme. The carboxylation involves at least four different steps and three transition states: the enolization of RuBP (3), the carboxylation of the 2,3-enediolate (4) and hydration of the resulting ketone (5), the carbon-carbon scission (6), and the stereospecific re-protonation of the resulting carboxylate of a 3PGA product (7).

Two molecules of CO_2 are produced by RubisCO's reaction with CO_2, whereas the competing reaction with O_2 forms one molecule of 3PGA and one molecule of 2PGA [4]. The latter one enters the photorespiratory carbon oxidation cycle, leading to a net loss of assimilated CO_2 and further on, the release of NH_3 and a considerable consumption of energy. The oxygenation reaction of RuBP in the CO_2-concentrating mechanisms present in cyanobacteria, algae, C_4 and CAM (Crassulacean Acid Metabolism) plants is efficiently decreased in these organisms, leading to a smaller proportion of photorespiration in relation to net photosynthesis.

RubisCO is characterized by a relatively slow catalytic turnover rate, k_{cat}. For this reason, large amounts are required to keep an adequate photosynthetic rate (e.g., [12]). Initial approaches focusing on photosynthetic CO_2-fixation improvements tried to identify or engineer a RubisCO enzyme with higher CO_2-specificities and/or higher catalytic rates [13]. These efforts had only limited success due to the fact that RubisCO is trapped in a constant trade-off between specificity and activity: usually, higher specificity for CO_2 results in a lower enzyme activity and vice versa. Although the emergence of the enzyme's carboxylation and oxygenation function is

used in the industrial Rectisol process [20]. Additionally, in aqueous solvents, different CO_2 hydration products are present, depending on the pH e.g., carbonic acid or carbonate.

In 1978, Halmann reported the first approach to reduce CO_2 on a semiconductor photoelectrode using p-GaP [21]. His analysis of the electrolyte solution showed the presence of formic acid, formaldehyde and methanol. Here, photoreduction of CO_2 was achieved using metal coated p-InP electrodes in non-aqueous solvents [20] and metal coated p-Si [22] and p-GaP electrodes [23, 24]. Furthermore, stepped (100) surfaces, such as Cu (911) and Cu (711), are able to convert CO_2 into longer chain hydrocarbons $C_{\geq 2}$ products with efficiencies approaching 80% [25]. Recently, also organic additives such as ionic liquids or pyridines have been employed to affect the product selectivity, which is a major obstacle in CO_2 reduction catalysis [26, 27]. Han et al. [28] reported on the ability of tuning the selectivity of electrochemical CO2 reduction on polycrystalline copper by N-substituted arylpyridinium additives for $C_{\geq 2}$. The selective, light-driven conversion of CO_2 to methanol at a p-GaP semiconductor electrode using a homogeneous pyridinium ion catalyst was described by Barton et al. [29]. Nevertheless, semiconductor cathodes such as n-GaAs and p-InP have also been demonstrated to convert CO_2 to CH_3OH without the addition of pyridine when biased to potentials more negative than -1 V versus SCE [27, 28]. In the presence of pyridine, however, the overpotential is reduced to ~ -0.2 V [29]. Quantum chemical calculations have been used in order to understand the catalytic role of pyridine [27]. The key lies in the homogenous chemistry of the 1,2-dihydropyridine/pyridine redox couple, driven by a dearomatization-aromatization process. Here, the 1,2-dihydropyridine acts as a recyclable organo-hydride, reducing CO_2 to CH_3OH via three hydride and proton transfer steps (Fig. 6.3): firstly, pyridine (Py) undergoes a H^+ transfer to form PyH^+, followed by an electron transfer step which forms pyridinium (PyH^0). Subsequently, the catalytic species 1,2 dihydropyridine (PyH_2) is formed via successive $1H^+/1e^-$ transfer. CO_2 reduction to CH_3OH and H_2O proceeds then homogeneously through three hydride and proton transfer steps. The novelty of the pyridine-catalysed reduction of CO_2 is its new approach in obtaining highly reduced species besides traditionally employed transition-metal-based electrocatalysts. Interestingly, the PyH_2/Py redox couple shows strong relation to the NADPH/$NADP^+$ couple in nature: both are catalytic hydride donors which use dearomatization to store energy, which is subsequently used to drive a proton transfer reaction.

Metal-organic species show a great versatility, insured by the possibility to modify the metal center and organic ligands. This fact turns them into interesting candidates for the development of selective catalysts. Metal-organic complexes such as macrocyclic complexes, phosphine complexes and polypyridyl complexes are particularly of interest, since they possess multiple potential oxidation states which can undergo multi-electron-transfer reactions. Among the macrocyclic complexes, one of the most studied catalysts is Ni(cyclam), Fig. 6.4a, which belongs to the class of tetraazamacrocycles. Within metalloamacrocycles, porphyrin-like materials with Ni, Fe or Co metallic centers also play an important

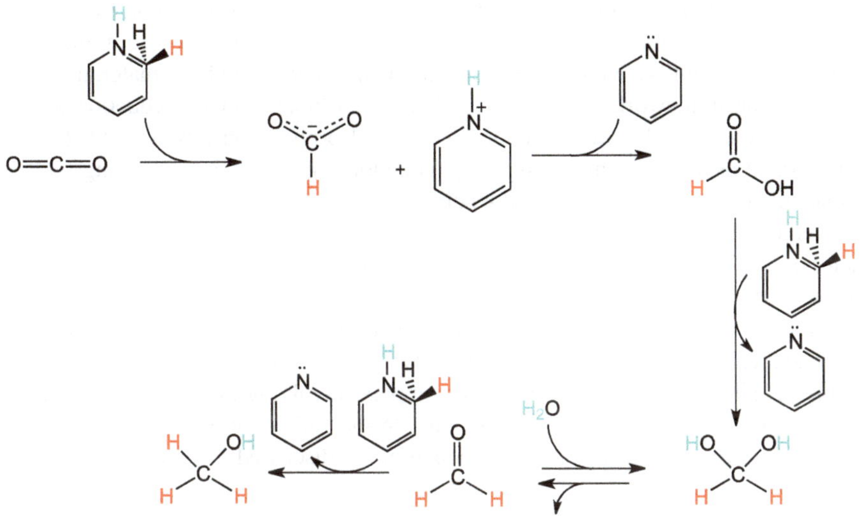

Fig. 6.3 Pyridine-catalysed reduction of CO_2 to methanol according to Lim et al. [27]

Fig. 6.4 Structures of typical molecular catalysts for CO_2 reduction. **a** $[Ni(cyclam)]^{2+}$; **b** metal porphyrin, **c** polypiridyl complex fac-$[Re](bpy)(CO)_3X]^{n+}$; **d** Dubois tridentate phosphine catalyst. According to Passalacqua et al. [16]

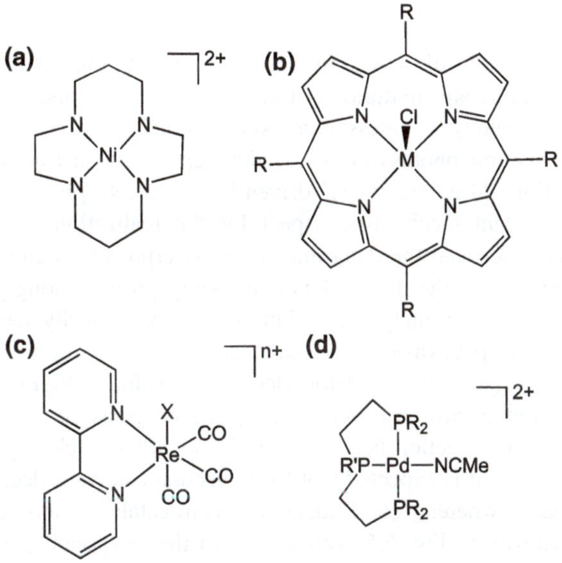

role (Fig. 6.4b). Polypyridine compounds which have been investigated for CO_2 reduction activity are mainly based on Ru, Os, Ir or Re metal centers. One of the most studied ones is $Re(bpy)CO_3X$ (bpy = 2,2'-bipyridine, X = Cl or Br, (**c**)). In phosphine compounds, Pd is commonly used as a metal center (**d**). Generally, higher product selectivity is achieved with bipyridine ligands which also require lower overpotentials. A limitation to their applicability, however, results from their low stability, a low turnover frequency (TOF) and their preferential use in organic media [16]. Recently, another class of inorganic-organic hybrid materials, the metal-organic frameworks (MOFS) have been used linked to CO_2 reduction (e.g., [30, 31]), especially, because of their excellent adsorption capacity for CO_2 [32].

Important aspects for improving CO_2 reduction kinetics and suppressing undesirable side reactions are factors influencing the overall efficiency of photocatalytic CO_2 reduction such as adsorption and activation of CO_2 (especially on semiconductors). Systems with sufficient solar energy conversion efficiencies to synthesise carbon-based fuels, demonstrating also long-term stability have not been realized for commercialization yet. Among the approaches to directly convert solar energy into chemical energy in the form of solar fuels, water-splitting is the most promising, since CO_2 reduction is more complex, involving multiple electron and proton transfer steps in the redox reactions. In the following, the proton-coupled electron transfer mechanisms of CO_2 reduction are discussed in more detail.

6.2.1 Sequential Versus Concerted Proton-Coupled Electron Transfer Mechanisms

As already discussed in Sect. 4.2, proton-coupled electron transfer (PCET) half reactions are ubiquitous in energy conversion and storage reactions in chemistry and biology. Quintessential examples include carbohydrates formation by RubisCO in natural photosynthesis, which represents an astounding example of PCET in action: 24 e^- and 24 H^+ driven by at least 48 photons are transferred. PCET is an important mechanistic aspect for the realization of (photo)electrochemical CO_2 reduction and the determination of overpotentials and product selectivity by computation of the thermodynamic energy profiles along potential reaction pathways. These first-principle calculations do not typically deal with the fact that PCET follows pathways where the electron and the proton are either transferred sequentially (sequential proton-electron transfer, SPET) or concertedly (concerted proton-electron transfer, CPET [33]) as discussed in Sect. 4.2. It is rather assumed that the selection between CPET and SPET is closely related to the nature of the catalyst: it is expected that for molecular catalysts, decoupled ET and PT steps take place, whereas solid metallic electrocatalysts undergo SPET steps: the CPET pathway in Fig. 6.5 corresponds to the 'diagonal' path and SPET pathways correspond to 'off-diagonal paths'. In order to address the selectivity between the pathways, the relative kinetics and thermodynamics have to be considered. Recently, a model was proposed which provides analytical expressions for the activation energies of the ET, PT and CPET steps and describes the transition between SPET and CPET. This model allows distinguishing when a certain pathway CPET or SPET is preferred over the other due to a lower activation barrier, since this preference depends on the relative values of the thermodynamic quantities

Fig. 6.5 Pathways for proton-coupled electron transfer (ET: electron transfer, PT = proton transfer) and the thermodynamic quantities which are associated with the reaction steps. According to Göttle and Koper [33]

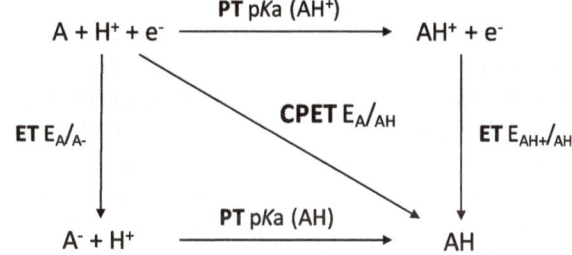

and reorganization energies [34]. A Marcus-type expression gives the rate constants of the ET, PT and CPET steps with the assumption that outer sphere charge transfer takes place [33]:

$$k_{ET} = k_{ET}^0 \exp\left(-\frac{(\lambda_{ET} + \Delta G_{ET})^2}{4\lambda_{ET}RT}\right)$$ (6.7)

$$k_{PT} = k_{PT}^0 \exp\left(-\frac{(\lambda_{PT} + \Delta G_{PT})^2}{4\lambda_{PT}RT}\right)$$ (6.8)

$$k_{CPET} = k_{CPET}^0 \exp\left(-\frac{(\lambda_{CPET} + \Delta G_{CPET})^2}{4\lambda_{CPET}RT}\right)$$ (6.9)

Here, k^0 parameters are pre-exponential factors, λ variables are the reorganization energies and ΔG values are the free reaction energies of the separate ET, PT and CPET reactions. The expressions can clearly distinguish between the impact of activation-related parameters (λ) values and thermodynamics-related parameters (ΔG). pH also influences the kinetics of the three reaction steps ET, PT and CPET since their thermodynamics scale differently with pH. This is illustrated in Fig. 6.6,

Fig. 6.6 Pourbaix diagram showing the thermodynamic equilibria of the ET (yellow/red, A + e$^-$ → AH), PT (blue, A$^-$ + H$^+$ → AH) and CPET (green, A + H$^+$ + e$^-$ → AH) reaction. According to Göttle and Koper [33]

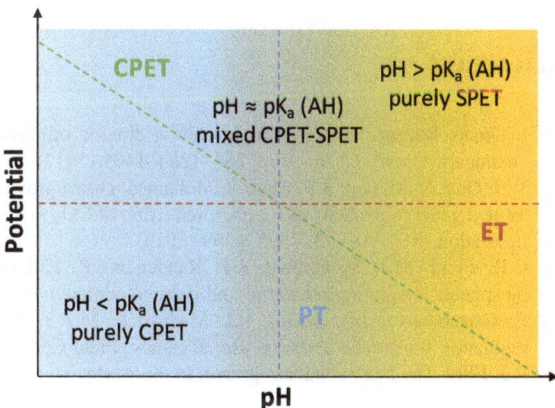

showing the thermodynamic equilibria of the steps in a Pourbaix diagram: when pH = pK$_a$(AH), all steps are equilibrated and the reorganization energies describe the competition between CPET and SPET. Since the thermodynamics and kinetics of the PT and CPET steps are pH sensitive, i.e. their rate increases (decreases) when the pH decreases (increases). In contrary, the thermodynamics of the ET reaction are not pH sensitive. As a result, the pH can influence the competition between CPET and SPET in case of reduction reactions (6.7)–(6.9).

Since evidence exists for the importance of SPET on metallic electrocatalysts (e.g., [35, 36], a complete picture has to consider both, sequential and concerted PCET steps. The community of heterogeneous catalysis typically employs the so-called computational hydrogen electrode (CHE), which was developed by Nørskov and co-workers [37]. The CHE method can, however, not account for SPET pathways. Recently, the introduction of a method by Göttle and Koper [33] which can be applied to any molecular or metallic electrocatalyst for the systematic prediction of the selectivity between SPET and CPET, allows the calculation of reaction schemes beyond the CHE methodology. This was demonstrated along the elucidation of the carboxylate adduct formation mechanism of a model molecular cobalt porphyrin catalyst which accounts for the possible coupling or decoupling of PT and ET in the initial stages of the electrocatalytic CO$_2$ reduction reaction.

As research efforts of artificial photosynthesis have evolved from studying isolated compounds to the construction of subsystems and devices, it has become evident that there is no single solution available yet solving our energy and environmental problems. It will further on take the cooperation of various forms of knowledge of various scientific disciplines to master the process of solar-driven water-splitting device incorporating CO$_2$ reduction and the generation of solar fuels. Significant contributions are made from catalysis, studies of nanomaterials and energy transfer dynamics, as well as from a more complete understanding of the mechanisms and principles governing the corresponding natural processes. The ability of biological systems of repairing, reproducing and evolving themselves extend their warranty to approximately 3 billion years and counting—given this perspective, there is still much to learn from nature.

References

1. I. Omae, Recent developments in carbon dioxide utilization for the production of organic chemicals. Coord. Chem. Rev. **256**, 1384–1405 (2012)
2. W.J. Ong, M.M. Gui, S.P. Chai, R. Mohamed, Direct growth of carbon nanotubes on Ni/TiO$_2$ as next generation catalysts for photoreduction of CO$_2$ to methane by water under visible light irradiation. RSC Adv. **3**, 4505–4509 (2013)
3. C.B. Field, M.J. Behrenfeld, J.T. Randerson, P. Falkowski, Primary production of the biosphere: integrating terrestrial and oceanic compounds. Science **281**, 237–240 (1998)
4. E. Carmo-Silva, J.C. Scales, P.J. Madgwick, M.A.J. Parry, Optimizing Rubisco and its regulation for greater resource use efficiency. Plant Cell Environ. **38**, 1817–1832 (2015)
5. R.J. Ellis, The most abundant protein in the world. Trends. Biochem. Sci. **4**, 241–244 (1979)

6. J.A. Raven, Rubisco: still the most abundant protein on Earth? New Phytol. **198**, 1–3 (2013)
7. B.J. Walker, A. van Loocke, C.J. Bernacchi, D.R. Ort, The costs of photorespiration to food production now and in the future. Annu. Rev. Plant Biol. **67**, 107–129 (2016)
8. I. Andersson, Catalysis and regulation in Rubisco. J. Exp. Bot. **59**(7), 1555–1568 (2008)
9. I. Andersson, T.C. Taylor, Structural framework for catalysis and regulation in ribulose-1,5-bisphosphate carboxylase/oxygenase. Arch. Biochem. Biophys. **414**, 130–140 (2003)
10. S. Knight, I. Andersson, C.I. Brändén, Crystallographic analysis of ribulose-1,5-diphosphate carboxylase from spinach at 2.4 Å resolution: subunit interactions and the active site. Plant Physiol. **54**, 678–685 (1990)
11. G. Tcherkez, Modelling the reaction mechanism of ribulose-1,5-bisphosphate carboxylase/ oxygenase and consequences for kinetic parameters. Plant Cell Environ. **36**, 1586–1596 (2013)
12. D. Mc Nevin, S. von Caemmerer, G.D. Farquhar, Determining Rubisco activation kinetics and other rate and equilibrium constants by simultanous multiple non-linear regression of a kinetic model. J. Exp. Bot. **57**, 3883–3900 (2006)
13. T.J. Erb, J. Zarzycki, Biochemical and synthetic biology approaches to improve photosynthetic CO_2 fixation. Curr. Opin. Chem. Biol. **34**, 72–79 (2016)
14. T.C. Taylor, I. Andersson, The structure of the complex between RubisCO and its natural substrate Ribulose-1,5-bisphosphate. J. Mol. Biol. **265**, 432–444 (1997)
15. T.J. Erb, B.S. Evans, K. Cho, B.P. Warlick, J. Sriram, B.M. Wood, H.J. Imker, J.V. Sweedler, F.R. Tabita, J.A. Gerit, A RubisCO-like protein links SAM metabolism with isoprenoid biosynthesis. Nat. Chem. Biol. **8**, 926–932 (2012)
16. R. Passalacqua, S. Perathoner, G. Centi, Semiconductor, molecular and hybrid systems for photoelectrochemical solar fuel production. J. Energ. Chem. **26**, 219–240 (2017)
17. B. Kumar, M. Llorente, J. Froehlich, T. Dang, A. Sathrum, C.P. Kubiak, Photochemical and photoelectrochemical reduction of CO_2. Annu. Rev. Phys. Chem. **63**, 541–569 (2012)
18. H. Ono, A. Yokosuka, T. Tasiro, H. Morisaki, S. Yugo, Characterization of diamond-coated Si electrodes for photoelectrochemical reduction of CO_2. New Diam. Front. Carbon Technol. **12**, 141–144 (2002)
19. K. Hirota, D.A. Tryk, T. Yamamoto, K. Hashimoto, M. Okawa, A. Fujishima, Photoelectrochemical reduction reduction of CO_2 in high-pressure CO_2+ methanol medium at p-type semiconductor electrodes. J. Phys. Chem. B **102**, 9834–9843 (1998)
20. S. Kaneco, H. Katsmumata, T. Suzuki, K. Ohta, Photoelectrochemical reduction of carbon dioxide at p-type gallium arsenide and p-type indium phosphide electrodes in methanol. Chem. Eng. J. **116**, 227–231 (2006)
21. M. Halmann, Photoelectrochemical reduction of aqueous carbon dioxide on p-type gallium phosphide in liquid junction solar cells. Nature **275**, 115–116 (1978)
22. Y. Nakamura, R. Hinogami, S. Yae, Y. Nakato. (1998). Photoelectrochemical reduction of CO_2 at a metal-particle modified p-Si electrode in non-aqueous solutions, in *Studies in Surface Science and Catalysis*, ed. by T. Makisy, T. Inui, T. Yamaguchi, vol 114, 565–568
23. S. Ikeda, M. Yoshida, K. Ito, Photoelectrochemical reduction products of carbon dioxide at metal coated p-GaP photocathodes in aqueous electrolytes. Bull. Chem. Soc. Jpn. **58**, 1353–1357 (1985)
24. S. Ikeda, Y. Saito, M. Yoshida, H. Noda, M. Maeda, K. Ito, Photoelectrochemical reduction products of carbon dioxide at metal coated p-GaP photocathodes in non-aqueous electrolytes. J. Electroanalyt. Chem. Interfac. Electrochem. **260**, 335–345 (1989)
25. Y. Hori, I. Takahashi, O. Koga, N. Hoshi, Selective formation of C2 compounds from electrochemical reduction of CO_2 at a series of copper single crystal electrodes. J. Phys. Chem. B **106**, 15–17 (2002)
26. B.A. Rosen, A. Salehi-Khojon, M.R. Thorson, W. Zhu, D.T. Whipple, P.J. Kenis, R.I. Masel, Ionic liquid-mediated selective conversion of CO_2 to CO at low overpotentials. Science **334**, 643–644 (2011)

27. C.H. Lim, A.M. Holder, J.T. Hynes, C.B. Musgave, Reduction of CO_2 to methanol catalysed by a biomimetic organo-hydride produced from pyridine. J. Am. Chem. Soc. **136**, 16081–16095 (2014)
28. Z. Han, R. Kortlever, H.-Y. Chen, J.C. Peters, T. Agapie, CO_2 reduction selective for C $\geq$ 2 products on polycrystalline copper with N-substituted pyridinium additives. ACS ent. Sci. **3**, 853–859 (2017)
29. E.E. Barton, D.M. Rampulla, A.B. Bocarsly, Selective solar-driven reduction of CO_2 to methanol using a catalysed p-GaP based photoelectrochemical cell. J. Am. Chem. Sc. **130**, 6342–6344 (2008)
30. C.G. Silva, A. Corma, H. Gracia, Metal-organic frameworks as semiconductors. J. Mater. Chem. **20**, 3141–3156 (2010)
31. C. Wang, Z. Xie, K.E. deKrafft, W. Lin, Doping metal-organic frameworks for water oxidation carbon dioxide reduction, and organic photocatalysis. J. Am. Chem. Soc. **133**, 13454–13455 (2011)
32. R.J. Li, R.J. Kuppler, H.C. Zhou, Selective gas adsorption and separation in metal-organic frameworks. Chem. Soc. Rev. **38**, 1477–1504 (2009)
33. A.J. Göttle, M.T.M. Koper, Proton-coupled electron transfer in the electrocatalysis of CO_2 reduction: prediction of sequential vs. concerted pathways using DFT. Chem. Sci. **8**, 458–465 (2017)
34. M.T.M. Koper, Theory of the transition from sequential to concerted electrochemical proton-electron transfer. Phys. Chem. Chem. Phys. **15**(5), 1399–1407 (2013)
35. S.C.S. Lai, S.E.F Kleijn, F.T.Z. Ozturk, V.C. van R. Vellinga, J. Koning, P. Rodriguez, M.T. M. Koper (2010). Effects of electrolyte pH and composition on the ethanol electro-oxidation reaction. Catal. Today 154(1–2), 92–104
36. P. Rodriguez, Y. Kwon, M.T.M. Koper, The promoting effect of adsorbed carbon monoxide on the oxidation of alcohols on a gold catalyst. Nat. Chem. **4**(3), 177–182 (2011)
37. J.K. Nørskov, J. Rossmeisl, A. Logadottir, L. Lindqvist, J.R. Kitchin, T. Bligaard, H. Jónsson, Origin of the overpotential for oxygen reduction at a fuel-cell cathode. J. Phys. Chem. B **108**, 17886–17892 (2004)

Chapter 7
Protection Mechanisms

In natural photosynthesis, the energy-converting enzymes Photosystem I and Photosystem II perform productive reactions efficiently despite the involvement of high energy intermediates in their catalytic cycles. This is achieved by kinetic control: forward reactions are faster than competing, energy-wasteful reactions due to appropriate cofactor spacing, driving forces and reorganizational energies. Despite that these high energy intermediates are short-lived, they have a tendency to react with oxygen which is produced during the water splitting reaction to form reactive oxygen species (ROS). While the production of ROS can be important in several cellular processes (e.g., defence against infection, cellular signalling), the presence of ROS is more often associated with damage to cellular components such as proteins, lipids and nucleic acids. To increase the efficiency of photosynthesis, several protective mechanisms dealing with oxygen have been developed by the photosynthetic reaction centres in order to minimise damage from reactive derivatives and provide the organism with a better chance of survival. These protective mechanisms involve fine-tuning of reduction potentials, switching of pathways and usage of short circuits, back-reactions and side-paths, all of which compromise at the same time efficiency. In dispersed semiconductor photocatalytic processes, oxygen plays a major role as well, acting as an electron acceptor of the electron promoted in the conduction band by light irradiation of the semiconductor surface: O_2 scavenges the photogenerated electrons and lowers the efficiency of the photocatalytic reaction. This chapter aims at summarizing protection mechanisms against O_2 which exists in natural photosynthesis and could be applied in artificial mimics, it discusses mechanisms of photoprotection in natural photosynthesis and outlines recent attempts to protect photoanodes and -cathodes against so-called 'photocorrosion', a process in which the photogenerated electron-hole pair induces a redox-decomposition of the semiconductor.

© Springer International Publishing AG, part of Springer Nature 2018
K. Brinkert, *Energy Conversion in Natural and Artificial Photosynthesis*,
Springer Series in Chemical Physics 117,
https://doi.org/10.1007/978-3-319-77980-5_7

7.1 Photodamage in Photosystem II

The ultimate energy source for photosynthesis is light, which is highly energetic and also potentially dangerous. The light induced decline of the photosynthetic activity is referred to as photoinhibition [1]. The major site of photoinhibition in the photosynthetic apparatus is the Photosystem II complex, whose electron transport is inhibited and the protein structure is damaged as a consequence of light exposure [2]. The molecular background of light sensitivity of PSII has turned out to be a complex story and the exact mechanisms are not fully understood yet. This is due to the complexity of events underlying the photoinhibitory phenomenon. It is clear that the main mechanisms which directly induce or lead to photodamage are the production of ROS (radical oxygen species, $^1O_2$, $O_2^{-\bullet}$, $OH^{-\bullet}$, H_2O_2) and the inactivation of the Mn_4CaO_5 cluster, leading to the formation of oxidized radicals ($Tyr_Z^{+\bullet}$, $P680^{+\bullet}$). The latter one is the most important target of UV-B (280–315 nm) light, whereas the primary and secondary quinone electron acceptor, Q_A and Q_B, and the tyrosine donors, TyrD and TyrZ, are also damaged [3]. The exact mechanism of the UV-induced impairment of the Mn cluster is not clear yet, but it is suggested to fully inhibit the S state cycle of water oxidation [1]. It is likely that the absorption of UV light by the high valence states of the Mn leads to the breakup of a bridging ligand between two Mn ions as it occurs in the model compound [4] which is supported by the observation that the Mn cluster is damaged most in the S_2 and S_3 state in which the Mn ions are in the Mn(III) and Mn(IV) states. Recent work by Zavafer et al. (2015) reported that subsequently to the photodamage of the Mn cluster, the PSII reaction center is further damaged by the light energy absorbed by photosynthetic pigments due to the limitations of electrons to the PSII reaction center [5].

In contrast to the widely accepted view about the primary role of the Mn_4CaO_5 cluster in sensitizing UV-induced photodamage of PSII, the situation is much more complex in case of visible light (400–700 nm). Here, modified or impaired function of the Q_A and Q_B acceptors, the inactivation of the Mn cluster and the production of various singlet oxygen species, especially, of singlet oxygen via Chl triplet formation are implicated. A large amount of available data in the literature demonstrates that a series of light-induced modifications take place at the acceptor side of PSII under conditions of strong illumination [1]. A conformational change has been shown to occur in *Chlamydomonas* cells which slows down the $Q_A^{-\bullet}$ to Q_B electron transfer rate, leading to an irreversible change of the D1 protein [6]. The central D1 protein is therefore also known as the most rapidly turned-over protein in the thylakoid membrane [7].

Excess excitation leads to the reduction of the quinone pool, in which the Q_B binding site becomes unoccupied due to the lack of reducible quinone molecules, leading to a destabilization of $Q_A^{-\bullet}$. Under strong reducing conditions, a double reduction and protonation of Q_A can occur, which is followed by the release of Q_AH_2 from the binding site [8]. This blocks the forward electron transport permanently, facilitating the formation of triplet excited state of P680 via charge recombination of the (^{3}[P680$^{+\bullet}$Pheo$^{-\bullet}$] → ^{3}P680) state. The further interaction of ^{3}P680 with O_2 leads to the formation of highly reactive singlet oxygen which damages its protein environment and leads to the actual inhibition of PSII electron transport [1]. Another way to produce highly reactive singlet oxygen is via the formation of the triplet excited state of chlorophylls. This process may occur via intersystem crossing from the singlet excited state of antenna chlorophyll (i.e., ^{1}Chl → ^{3}Chl), although the chlorophylls in the light harvesting systems are in general protected against ^{3}Chl formation by carotenoids, which also act at $^1O_2$ quenchers [9].

7.1.1 Protection Against Light Damage

Oxygenic photosynthetic organisms possess several photoprotection mechanisms which help to eliminate the harmful effects of light [10]. The non-photochemical quenching mechanism (NPQ) dissipate absorbed light energy in the antenna system before it can reach the Photosystem II reaction centre [11]. These mechanisms include the LHCII dependent NPQ, which depends on the light induced lumen acidification. The NPQ mechanisms help to decrease the excitation pressure on PSII and provide protection against electron transport dependent and/or triplet chlorophyll dependent photodamage. Another way to avoid photodamage is the elimination of potentially harmful radical states, especially, ^{3}[P680$^{+\bullet}$Pheo$^{-\bullet}$].

This can occur via non-radiative charge recombination (Fig. 7.1, P680 is here specified with P_{D1}) which can be modulated via redox potential changes of pheophytin or Q_A [1] (see also Sect. 4.3). The results of the photoprotective effect of the redox potential changes at the acceptor site can easily be explained by the decreased efficiency of thermally activated back reaction of P680$^{+\bullet}Q_A^{-\bullet}$ to ^{3}[P680$^{+\bullet}$Pheo$^{-\bullet}$], which depends on the redox gap between Pheo$_{D1}$ and Q_A and also by an enhanced non-radiative charge recombination from P680$^{+\bullet}Q_A^{-\bullet}$ and ^{1}[P680$^{+\bullet}$Pheo$^{-\bullet}$] which also competes with ^{3}[P680$^{+\bullet}$Pheo$^{-\bullet}$] and ^{3}P680 formation [12].

7.1.1.1 Protection Mechanisms Against Oxygen

The by-product of photosynthetic water-splitting in chloroplasts is the production of molecular oxygen, which is essential for cellular respiration in mitochondria, a process, which is required for all aerobic organisms. When solar energy absorption by chlorophylls exceeds its utilization, reactive oxygen species are formed. Singlet oxygen ($^1O_2$) is generated by the triplet-singlet excitation energy transfer from the triplet chlorophyll to molecular oxygen due to intersystem crossing (see Sect. 7.1.1), by charge recombination from the singlet chlorophyll in the antenna complex or by the charge recombination of the primary radical pair ^{1}[P680$^{+\bullet}$Pheo$^{-\bullet}$] in the reaction centre [13]. The latter one is considered as the main reaction pathway for $^1O_2$ generation in PSII. Singlet oxygen scavenging occurs either by excitation

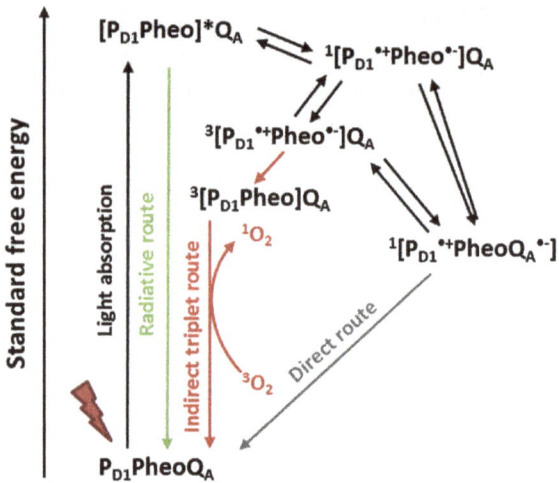

Fig. 7.1 The two charge recombination pathways in Photosystem II for $P^{•+}Q_A^{•-}$ and the size of the energy gap between Q_A and $Pheo_{D1}$ determines the back-reaction rate and the recombination route: the indirect route via $P_{D1}^{•+}Pheo_{D1}^{•-}$, which then decays to the ^3P triplet state. This lies about 1.3 eV above the ground state and can easily promote the triplet to singlet oxygen formation. The direct route is favoured, when the energy gap between $Pheo_{D1}$ and Q_A is sufficiently large (presumably during photoactivation). Then $^1O_2$ formation is completely avoided. According to Sugiura et al. (2014)

energy transfer of by electron transport. The first one is accomplished by carotenoids in the antenna systems. It results in the formation of the ground triplet state of molecular oxygen and the triplet excited state of the carotenoid (Car):

$$^1O_2 + Car \rightarrow O_2 + {}^3Car^* \tag{8.1}$$

$$^3Car^* \rightarrow Car + heat \tag{8.2}$$

In LHCII, singlet oxygen is effectively quenched by xanthophylls: α-xanthophyll (lutein) is an efficient quencher of triplet chlorophyll, whereas β-xanthophyll (zeaxanthin and neoxanthin) serve as a quencher of $^1O_2$ [14]. Under high light conditions, violaxanthin is also enzymatically converted to zeaxanthin. In the PSII reaction centre and the core antenna complex, β-carotin quenches the formed $^1O_2$.

Enzymatic scavenging of $^1O_2$ is also carried out in the photosynthetic reaction centre. Here, $cytb_{559}$ has been reported to possess superoxide dismutase (SOD) activity and provides the first line of defence against $^1O_2$ in the membrane interior (Fig. 7.2): the redox active heme iron of $cytb_{559}$ has been shown to be both oxidized and reduced by $^1O_2$, depending on the oxidation state of the heme iron. In intact PSII, most of the $cytb_{559}$ is in a high potential (HP) form (E_m = +310 to +410 mV), which can be readily converted into the intermediary (IP, E_m = +125 to +240 mV) or low potential form (LP, E_m = −40 to +80 mV [12]). Tiwari and

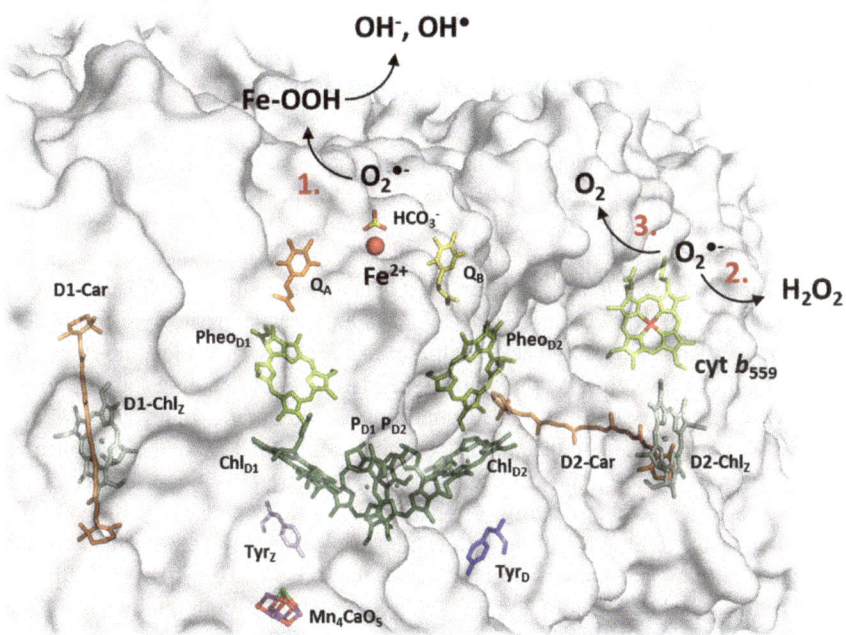

Fig. 7.2 The arrangement of cofactors in PSII according to the 1.95 Å crystal structure (PDB reference 4UB6) with the metal centers potentially involved in the reduction of superoxide. (1) Reduction of $O_2^{\cdot-}$ by the ferrous non-heme iron to OH^- and $OH^{\cdot}$ (or in a shortcut reaction to H_2O_2); (2) reduction of $O_2^{\cdot-}$ to O_2 by the ferrous iron of the LP form of cyt b_{559} and (3) of $O_2^{\cdot-}$ by the ferrous iron of the HP form of cyt b_{559} to H_2O_2

Pospíšil (15) proposed that the unprotonated IP form of cytb_{559} possess superoxide oxidase (SOO) activity, which catalyses the oxidation of $^1O_2$ to O_2 (E_m ($O_2/O_2^{\cdot-}$) = -160 mV, pH 7) [15]. Hereby, the ferric heme iron (Fe(III)) is reduced to ferrous heme iron (Fe(II)). The HP form is proposed to serve as a superoxide reductase (SOR) which catalyses the one-electron reduction of $^1O_2$ to H_2O_2 (E_m ($O_2^{\cdot-}/H_2O_2$) = $+890$ mV, pH 7).

The further reduction of the ferric-hydroperoxo species is suggested to result in the formation of $OH^{\cdot}$. In this reaction, the ferric iron is proposed to be reduced by an endogenous reductant, the most likely being Q_A^-, whereas the ferrous iron produced is suggested to reduce the hydroperoxo ligand [16]. Also the non-heme iron at the electron acceptor site has been proposed to exhibit SOD (superoxide dismutase) activity [16, 17], Fig. 7.2. In SOR, one-electron reduction of $O_2^{\cdot-}$ by the ferrous non-heme iron forms the ferric peroxo intermediate (Fe^{3+}-OO^-), which forms upon protonation the ferric-hydroperoxo species. Since the midpoint potential of the Fe^{3+}/Fe^{2+} redox couple of the non-heme iron is $E_m = +400$ mV, the reduction of bound peroxide by the non-heme iron is feasible.

7.2   Catalyst Stability and Photocorrosion in Photoelectrochemical Cells

7.2.1   Role and Function of Protection Layers

An optimized tandem junction solar-driven water-splitting system consists of semiconductors with band gaps of 1.6–1.8 eV for the photoanode and 0.95–1.2 eV for the photocathode, with the exact values depending on the electrocatalysts and the design of the water-splitting cell [18]. These requirements rule out most of the oxide semiconductors such as TiO_2 or $SrTiO_3$ and rather suggest the suitability of the technology important Group IV, III–V, II–VI and chalcopyrite semiconductors as light absorbers. The drawback of these materials is that they are typically unstable (either dissolving or developing insulating oxide coatings as discussed earlier) under aqueous HER and OER conditions. Two parallel paths are currently followed in order to increase the stability of the materials and make them applicable as well as photoelectrodes in PEC cells: (i) methods are developed to protect the otherwise unstable semiconductors to enable their use in efficient water-splitting cells (ii) new materials are discovered which are inherently stable under water-splitting conditions and possess a band gap in the range of 0.95–1.8 eV [19].

Important for the development of PEC materials for both pathways is the electrochemical environment in which the HER and OER reactions are carried out: to be efficient, a water-splitting cell has to make use of electrolytes which can support photocurrent densities of about 10 mA cm^{-2} under non-concentrated 1 sun illumination (AM 1.5). In the absence of external electrical inputs, this corresponds to solar-to-hydrogen conversion efficiencies of $\geq 12.3\%$ [20]. An inefficient ion transport will cause the development of a potential gradient in the cell, manifesting partly as a pH gradient in aqueous electrolytes as well as in a possible concentration gradient associated with the transport of other ionic species in the electrolyte. Many semiconductors which are capable of supporting the requisite current densities without developing significant pH gradients are chemically unstable in the required strongly acidic or alkaline electrolytes [21]. The strategy of coating unstable semiconductor surfaces with thermodynamically stable films (Fig. 7.3) provide a promising approach to overcome this obstacle of stability. The protective film should be electrically conductive, electrochemically stable, optically transparent and it should prevent the semiconductor surface from direct contact with the

Fig. 7.3 Schematic drawing of a protective film on the light absorbing semiconductor for photoelectrochemical fuel generation along with the required properties. According to Hu et al. (2015)

electrolyte [19]. As outlined by Hu et al. (2015), the composition of the protective film, the deposition technique and the interface structure can play a significant role in determining the resulting photoelectrocatalytic behaviour.

The chosen deposition technique influences the uniformity, stoichiometry, interfacial characteristics and other structural and electronic properties of the film and may also critically impact the stability and performance of the composite photoelectrode. Physical-vapor deposition (PVD) techniques e.g., thermal or electron beam evaporation, magnetron sputtering and pulsed laser deposition enable a precise control over the composition and thickness of the protective film. Drawbacks of PVD are the line-of-sight deposition which can result in shadowing and inhomogeneous deposition of the film on structured semiconductor surfaces and the lack of chemical control at the interface between the semiconductor and the deposited film. Chemical vapour deposition (CVD) techniques have also been used to deposit protection coatings. One obstacle is that these techniques are limited by the precursor volatility, stability and deposition chemistry, which limit the available deposition condition and material composition [19]. It can be used to coat non-planar surfaces under suitable deposition conditions. The subset of CVD techniques, atomic layer deposition (ALD), uses sequential, self-limiting surface reactions and is particularly suitable to deposit conformal and uniform high aspect-ratio or porous materials [22]. In comparison to traditional CVD and PVD techniques, ALD offers a high degree of thickness control relative to traditional CVD and PVD techniques and due to these advantages, it is the most frequently used technique to fabricate thin films for the protection of photoelectrodes.

Due to the earlier described influence of the band bending of a semiconductor photoelectrode by the charge transfer equilibrium between its Fermi level and the Fermi level (electrochemical potential) of the liquid contact, the deposition chemistry as well as the structure of the resulting semiconductor/protective film interface can significantly influence the electronic properties and energy-conversion efficiencies of the photoelectrode: the introduction of protective layers often converts a PEC cell into a photovoltaic cell or into a hybrid junction which incorporates aspects of a semiconductor/liquid junction and also aspects of a solid-state junction [19].

Photocorrosion of nonoxide photoanode materials in aqueous electrolytes has already been observed early [23] and significant attention has been devoted to protect them for the realisation of a stable water-splitting system. The corrosion stability of light absorbing semiconductors and oxide coatings are generally referenced in a Pourbaix diagram described earlier. Many metals form insoluble oxides and self-passivation layers under water oxidation conditions (E > 1.23 V vs. NHE) and certain pH conditions. Others form soluble species, resulting in film destruction or dissolution. Considering their insolubility over certain pH ranges and without considering their electronic transport properties, e.g., TiO_2, ZrO_2, SnO_2, NiO_x, FeO_x, MnO_x and WO_3 can be used for photoanode protective coatings [19]. The most prominent example is probably TiO_2, which has also been used as a gate dielectric [24]. Different depositions conditions produce significantly different conductivities and materials properties [25]. Hereby, the film thickness of TiO_2 protection layers is crucially important, since films thicker than a few nm are highly resistive [26]. Atomic layer deposition of TiO_2 films at 150 °C from tetrakis dimethylamido titanium (TDMAT) is highly resistive as deposited, but the films are very conductive toward anodic current flow following the sputter deposition of a Ni film or Ni islands [27]. Consequently, over 100 nm of ALD-TiO_2 can be deposited onto a semiconductor without concerns for resistive losses in the film.

Early demonstrations of PEC-based hydrogen evolution did not explicitly consider efforts to protect the semiconductor surface from corrosion since many photocathodes exhibit stability under HER conditions. Some materials are even cathodically protected under illumination due to the operation in the immunity region of the Pourbaix diagram. Other materials have extremely slow kinetics for reduction of surface species and undergo minimal bulk reduction processes [19]. Protection layers can, however, also increase the stability of the material and the complexity of the device structure and fabrication. In certain cases, some p-type semiconductors used as light absorbers for hydrogen evolution reactions may spontaneously form protective surface layers upon contact with an aqueous electrolyte. Depending on the semiconductor, these layers might be insulating as with Si (SiO_2) or may be n-type conducting such as InP (In_2O_3). Furthermore, the physical and electronic properties (crystallinity, defect density, conductivity, work function etc.) are also affected by the conditions under which the layer is formed. Therefore, a protection/passivation layer for photocathodes is highly material specific. p-Si photocathodes are not thermodynamically stable under aqueous conditions; in alkaline media, Si rapidly oxidizes and dissolves via chemical etching and the dissolution cannot be entirely ascribed to electrochemical means. Under neutral and acidic conditions, Si forms a passivating oxide which is stable toward dissolution, although, the oxide is a barrier for electron transfer (Hu et al. 2015). In 1977, Bard and coworkers [28] coated p-Si (and also p-GaAs) with TiO_2 grown by CVD with only small cathodic photocurrents being observed. Recently, conducting oxides have been used on p-Si photocathodes. One common approach which was also used by Seger et al. (2013) is to use a photovoltaic p-n^+ junction which contains a TCO overlayer [29]: they found that a 5 nm metallic Ti layer followed by a thick (>100 nm) layer of TiO_2 and Pt cocatalyst on top of a Si p-n^+ homojunction allowed operation of the electrically isolated photovoltaic buried junction device for several weeks of continuous hydrogen evolution in 1 M $HClO_{4(aq)}$. In the same way as Si, III–V materials (e.g., GaAs, GaP, InP) oxidize readily under aqueous conditions. The stability and electronic properties of the resulting oxides vary significantly with the composition of the semiconductor. At negative potentials, group III elements can also be further reduced to their metallic form [30], e.g., metallic Ga or In can be formed from GaAs, GaP or InP. Lee et al. (2012) showed that high surface area InP structures formed by reactive ion etching with 2–5 nm layers of ALD-deposited TiO_2 were able to generate photocurrent onset potentials of >600 mV versus RHE, current densities of 37 mA/cm^2 under 100 mW/cm^2 of simulated solar illumination and overall cell efficiencies of 13% [31].

Chemical stability, and protection are critically important for solar fuel systems, including device geometry, semiconductor/liquid junction performance, ion transport and overall system costs. Protective films can enable the use of nonoxide semiconductors in environments containing liquid electrolytes, not only for water-splitting, but also for photochemical processes such as hydrogen-halide electrolysis and electrochemical reduction of CO_2 (see Sect. 6.2). Several classes of potential protection materials such as metal carbides, metal nitrides and 2-dimensional materials such as graphene and MoS_2 and hexagonal boron nitride have not been fully explored yet and may also provide a wide variety of compositions for the protection of semiconducting photoelectrodes [19].

References

1. I. Vass, Molecular mechanisms of photodamage in Photosystem II. Biochim. Biophys. Acta **1817**, 209–217 (2012)
2. E.M. Aro, I. Virgin, B. Andersson, Photoinhibition of Photosystem II. Inactivation, protein damage and turnover. Biochim. Biophys. Acta **1143**, 113–134 (1993)
3. I. Vass, L. Sass, C. Spetea, A. Bakou, D. Ghanotakis, V. Petrouleas, UV-B induced inhibition of Photosystem II electron transport studied by EPR and chlorophyll fluorescence. Impairment of donor and acceptor site components. Biochemistry **35**, 8964–8973 (1996)
4. W.F. Ruettinger, M. Yagi, K. Wolf, S. Bernasek, G.C. Dismukes, O_2 evolution from the manganese-oxo cubane core $Mn_4O_4^{\delta+}$: a molecular mimic of the photosynthetic water oxidation enzyme? J. Am. Chem. Soc. **122**, 10353–10357 (2000)
5. A. Zavafer, M.H. Cheah, W. Hillier, W.S. Chow, S. Takahashi, Photodamage to the oxygen evolving complex of Photosystem II by visible light. Sci. Rep. **5**, 16363–16368 (2015)
6. I. Ohad, N. Adir, H. Koike, D.J. Kyle, Y. Inoue, Mechanism of photoinhibition in vivo. A reversible light-induced conformational change of reaction centre II is related to an irreversible modification of the D1 protein. J. Biol. Chem. **265**, 1972–1979 (1990)
7. M. Edelman, A.K. Mattoo, D1-protein dynamics in Photosystem II: the lingering enigma. Photosynth. Res. **98**, 609–620 (2008)
8. I. Vass, S. Styring, T. Hundal, A. Koivuniemi, E.-M. Aro, B. Andersson, Reversible and irreversible intermediates during photoinhibition of Photosystem II: stable reduced QA species promote chlorophyll triplet formation. Proc. Nat. Acad. Sci. USA **89**, 1408–1412 (1992)
9. H.A. Frank, R.J. Cogdell, Carotenoids in photosynthesis. Photochem. Photobiol. **63**, 257–264 (1996)
10. S. Takahashi, M.R. Badger, Photoprotection in plants: a new light on Photosystem II damage. Trends Plant Sci. **16**, 53–60 (2001)
11. P. Müller, X.P. Li, K.K. Niyogi, Non-photochemical quenching. A response to excess light energy. Plant Physiol. **125**, 1558–522 (2001)
12. M. Sugiura, C. Azami, K. Koyama, A.W. Rutherford, F. Rappaport, A. Boussac, Modification of the pheophytin redox potential in Thermosynechococcus elongatus Photosystem II with PsbA3 as D1. Biochim. Biophys. Acta Bioenerg. **1837**(1), 139–148 (2014)
13. P. Pospíšil, A. Arató, A. Krieger-Liszkay, A.W. Rutherford, Hydroxyl radical generation in Photosystem II. Biochemistry **43**, 6783–6792 (2004)
14. L. Dall'Osto, A. Fiore, S. Cazzangia, G. Giuliano, R. Bassi, Different roles of alpha- and beta-branch xanthophylls in photosystem assembly and photoprotection. J. Bio. Chem. **282**, 35056–35068 (2007)
15. A. Tiwari, P. Pospíšil, Superoxide oxidase and reductase activity of cytochrome b_{559} in Photosystem II. Biochim. Biophys. Acta **1787**, 985–994 (2009)
16. P. Pospíšil, A. Arató, A. Krieger-Liszkay, A.W. Rutherford, Hydroxyl radical generation in Photosystem II. Biochemistry **43**, 6783–6792 (2004)

17. J.H. Nugent, Photoreducible high spin iron electron paramagnetic resonance signals in dark-adapted Photosystem II: are they oxidised non-haem iron formed from interaction of oxygen with PSII electron acceptors? Biochim. Biophys. Acta **1504**, 288–298 (2001)
18. S. Hu, C. Xiang, S. Haussener, A.D. Berger, N.S. Lewis, An analysis of the optimal band gaps of light absorbers in integrated tandem photoelectrochemical water-splitting systems. Energy Environ. Sci. **6**, 2984–2993 (2013)
19. S. Hu, N.S. Lewis, J.W. Ager, J. Yang, J.R. McKone, N.C. Strandwitz, Thin-film materials for the protection of semiconducting photoelectrodes in solar-fuels generators. J. Phys. Chem. C **119**, 24201–24228 (2015)
20. R.H. Coridan, A.C. Nielander, S.A. Francis, M.T. McDowell, V. Dix, R.H. Chatman, N. Lewis, Methods for comparing the performance of energy conversion systems for use in solar fuels and solar electricity generation. Energy Environ. Sci. **8**, 2886–2901 (2015)
21. E. Hernandez-Pagan, N. Vargas-Barbosa, T. Wang, Y. Zhao, E.S. Smotkin, T.E. Mallouk, Resistance and polarization losses in aqueous buffer-membrane electrolytes for water-splitting photoelectrochemical cells. Energy Environ. Sci. **5**, 7582–7589 (2012)
22. S.M. George, Atomic layer deposition: an overview. Chem. Rev. **110**, 111–131 (2010)
23. H. Gerischer, On the stability of semiconductor electrodes against photodecomposition. J. Electroanal. Chem. Interfacial Electrochem. **82**, 133–143 (1977)
24. S.A. Campbell, H.S. Kim, D.C. Gilmer, B. He, T. Ma, W.L. Gladfelter, Titanium dioxide (TiO_2)-based gate insulators. IBM J. Res. Dev. **43**(9), 383–392 (1999)
25. M.F. Lichterman, K. Sun, S. Hu, X. Zhou, M.T. McDowell, M.R. Shaner, M.H. Richter, E. J. Crumlin, A.I. Carim, F.H. Saadi, B.S. Brunschwig, N.S. Lewis, Protection of inorganic semiconductors for sustained efficient photoelectrochemical water oxidation. Catal. Today **262**, 11–23 (2016)
26. Y.W. Chen, J.D. Prange, S. Dühnen, Y. Park, M. Gunji, C.E.D. Chidsey, P.C. McIntyre, Atomic layer-deposited tunnel oxide stabilizes silicon photoanodes for water oxidation. Nat. Mater. **10**(7), 539–544 (2011)
27. S. Hu, M.R. Shaner, J.A. Beardslee, M.F. Lichterman, B.S. Brunschwig, N.S. Lewis, Amorphous TiO_2 coatings stabilize Si, GaAs, and GaP photoanodes for efficient water oxidation. Science **344**(6187), 1005–1009 (2014)
28. P.A. Kohl, S.N. Frank, A.J. Bard, Semiconductor electrodes XI. Behaviour of n- and p-type single crystal semiconductors covered with thin TiO_2 films. J. Electrochem. Soc. **124**, 225–229 (1977)
29. B. Seger, T. Pedersen, A.B. Laursen, P.C.K. Vesborg, O. Hansen, I. Chorkendorff, Using TiO_2 as a conductive protective layer for photocathodic H_2 evolution. J. Am. Chem. Soc. **135**, 1057–1064 (2013)
30. B. Kaiser, D. Fertig, J. Ziegler, J. Klett, S. Hoch, W. Jaegermann, Solar hydrogen generation with wide-band gap semiconductors: GaP (100) photoelectrodes and surface modification. Chem. Phys. Chem. **13**, 3053–3060 (2012)
31. M.H. Lee, K. Takei, J. Zhang, R. Kapadia, M. Zheng, Y. Chen, J. Nah, T.S. Matthews, Y. Chueh, J.W. Ager, A. Javey, P-type InP nanopillar photocathodes for efficient solar-driven hydrogen production. Angew. Chem. Int. Ed. **51**, 10760–10764 (2012)

Chapter 8
Biomimetic Systems for Artificial Photosynthesis

The increased understanding of photosynthetic energy conversion and advances in chemical synthesis have made it possible to create artificial biomimetic systems and semibiological hybrids which are able to carry out parts of the natural photosynthetic process. These biomimetic systems reduce the complicated natural mechanism to its basic elements, leading to a better understanding of photosynthesis and to potential energy sources with its application in molecular-scale optoelectronics, photonics, sensor design and other areas of nanotechnology [1]. A variety of research groups have developed artificial reaction center molecules and mimics of nature's water-splitting catalyst, the Mn_4CaO_5 cluster. These systems contain a chromophore in analogy to chlorophyll e.g., a porphyrin, which is covalently linked to one or more electron acceptors such as fullerenes or quinones and secondary electron donors. The redox equivalents are spatially separated by electron transfer chains, reducing electronic coupling and slowing recombination of the charge separated state to the extend at which catalysts are able to use them for storing energy e.g., for fuel production [2]. Although, attempts of these approaches of realizing artificial photosynthesis fall short in efficiencies terms for practical solar fuel application, they demonstrate that solar fuel production is actually achievable in the laboratory. This chapter aims at outlining some of the synthetic, biomimetic highlights, developed to further the understanding of natural photosynthesis and to realize a stable, robust artificial system using earth abundant elements to split water and synthesise fuels.

8.1 Photosynthetic Model Systems: Concepts and Ideas to Realize Artificial Photosynthesis

Following the biological role model, an artificial photosynthetic system requires an antenna/reaction center complex, harvesting sunlight and generating an electrochemical potential. Since the late 1970s, molecule-based artificial reaction centers

© Springer International Publishing AG, part of Springer Nature 2018   97
K. Brinkert, *Energy Conversion in Natural and Artificial Photosynthesis*,
Springer Series in Chemical Physics 117,
https://doi.org/10.1007/978-3-319-77980-5_8

have been reported (e.g., [3, 4]). One example are studies of electron donor-acceptor dyads for photoinduced charge separation which have revealed much concerning the basic principles governing electron transfer. A temporal stability of the charge-separated state with a kinetic ability to carry out redox reaction with catalysts or other species require at least a three-component (triad) system. The carotenoid (C)–porphyrin (P)–fullerene (C_{60}) molecular triad in Fig. 8.1 illustrates such a system.

Excitation of the porphyrin in 2-methyltetrahydrofuran solution leads to the formation of its first excited singlet state C–P^1–C_{60} which decays by photoinduced electron transfer with a time constant τ of 32 s to the fullerene. This yields in C–$P^{\cdot-}$–$C_{60}^{\cdot-}$ with a quantum yield of 0.99. Although, the molecule has converted light into electrochemical potential, the charge recombination to the ground state rapidly occurs with $\tau = 3.3$ ns, wasting the stored energy as heat [2]. This short life time makes it difficult to determine the redox potential. The charge recombination reaction competes, however, in the triad molecule with a rapid transfer of the positive charge to the carotene (hole transfer $\tau = 125$ ps), resulting in the formation of $C^{\cdot-}$–P–$C_{60}^{\cdot-}$ with an overall quantum yield of 0.95. This charge separated state possess a life time of 57 ns. Related C–P–C_{60} systems even demonstrate life times of 170 ns at ambient temperatures and ~ 1 μs at 77 K [5]. Due to a sufficient thermodynamic driving force and electronic coupling between initial and final states, each electron transfer step occurs rapidly enough in order to compete with the loss of energy by other pathways. The charges in the final state $C^{\cdot-}$–P–$C_{60}^{\cdot-}$ are well-separated, resulting in a slow recombination although the driving force is sufficiently large (>1.0 eV). Natural photosynthetic reaction centers employ exactly the same strategy to achieve a long-lived, energetic charge-separated state in high yield: the application of sequential rapid, short-range electron transfers. This idea was firstly developed by Gust et al. in 1983 and many other examples of promising systems have followed until then [6].

Most sunlight used for photosynthesis is not absorbed by the reaction centers, but rather by antenna systems which transfer the resulting singlet excitation energy to reaction centers (Sect. 3.1.1). Universally used as the primary excited state electron

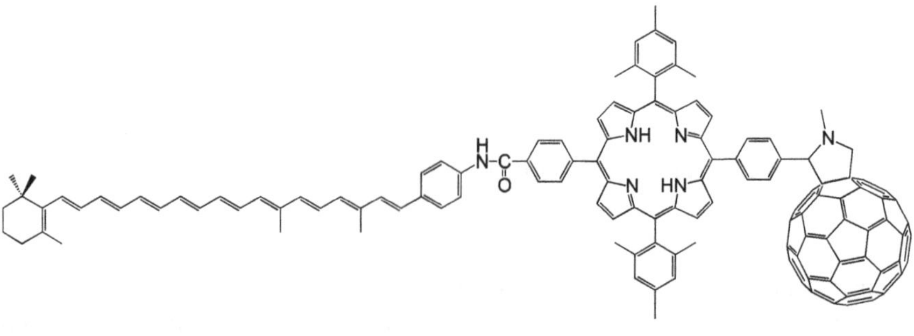

Fig. 8.1 Structure of a molecular triad consisting of a carotenoid–porphyrin–fullerene system representing an artificial reaction center. According to Gust et al. 2009

donors in the reaction centers are chlorophylls, although, e.g., chlorophyll a has relatively weak absorption bands in the visible region between ca. 430 and 660 nm. Accessory antenna chromophores are used to harvest energy in regions where chlorophyll absorption does not—or only to a weak extend—take place. Commonly, photosynthetic organisms employ carotenoid polyenes and in some cases phyco-erythrins and phycocyanins. Although, nature developed a wide variety of antenna morphologies and compositions in order to adjust to the wide variety of light con-ditions under which photosynthesis occurs, all antennas use multiple chromophores for broad absorption spectrum, rapid singlet-singlet energy transfer among the chromophores and rapid energy transfer to the reaction center chlorophylls.

Advances in the photosynthetic antennas and improvements in synthetic and spectroscopic methods have allowed the development and design of artificial light-harvesting antennas [7]. The linkage of synthetic porphyrins forms antenna arrays and photonic 'wires', gates and switches [8]. Kuciauskas et al. (1999) linked an artificial antenna array to an artificial reaction center for the formation of a functional unit [9]. The antenna complex of $(P_{ZP})_3-P_{ZC}-P-C_{60}$ hexad (Fig. 8.2) comprises four zinc tetraarylporphyrins $((P_{ZP})_3-P_{ZC})$, joined to a free base por-phyrin–fullerene artificial reaction center analogous to $P-C_{60}$ (see Fig. 8.1). Time-resolved spectroscopy revealed that excitation of any zinc porphyrin moiety (P_{ZP}) is followed by energy transfer of the central Zn porphyrin to yield $(P_{ZP})_3-{}^1P_{ZC}-P-C_{60}$ with a time constant of 50 ps. In 240 ps, the excitation is transferred to the free base porphyrin, which results in the formation of $(P_{ZP})_3-P_{ZC}-{}^1P-C_{60}$. Electron transfer to the fullerene was measured to occur with a time constant of 3 ps. The resulting $(P_{ZP})_3-P_{ZC}-P^{\bullet+}-C_{60}^{\bullet-}$ is formed with a quantum yield of 0.70 and a lifetime of 1.3 ns.

One requirement for photosynthesis is the manipulation of highly energetic redox carriers by reactive complex biomolecules. Therefore, it incorporates extensive photoprotective and regulatory mechanisms to limit photodamage (Sect. 7.1). Here, carotenoid polyenes quench precursor chlorophyll triplet states and provide therefore photoprotection from singlet oxygen damage. Furthermore, they participate in the non-photochemical quenching (NPQ) regulatory mechanism found in plants. Artificial photosynthetic systems also require protection and reg-ulation (see Chap. 7), which led to the closer investigation of the carotenoid role in artificial systems (e.g. [10]) and the self-downregulation in bright-light [11].

8.1.1 Electron Transfer Studies

Photosystem II comprises the essential function of combining single-photon exci-tation with the multielectron process of water oxidation. Megiatto et al. (2012) re-ported on the preparation and photophysical investigation of a bioinspired molecular triad engineered to functionally mimic the initial charge-separation events in PSII [12]. The three units (Fig. 8.3) were designed and covalently linked

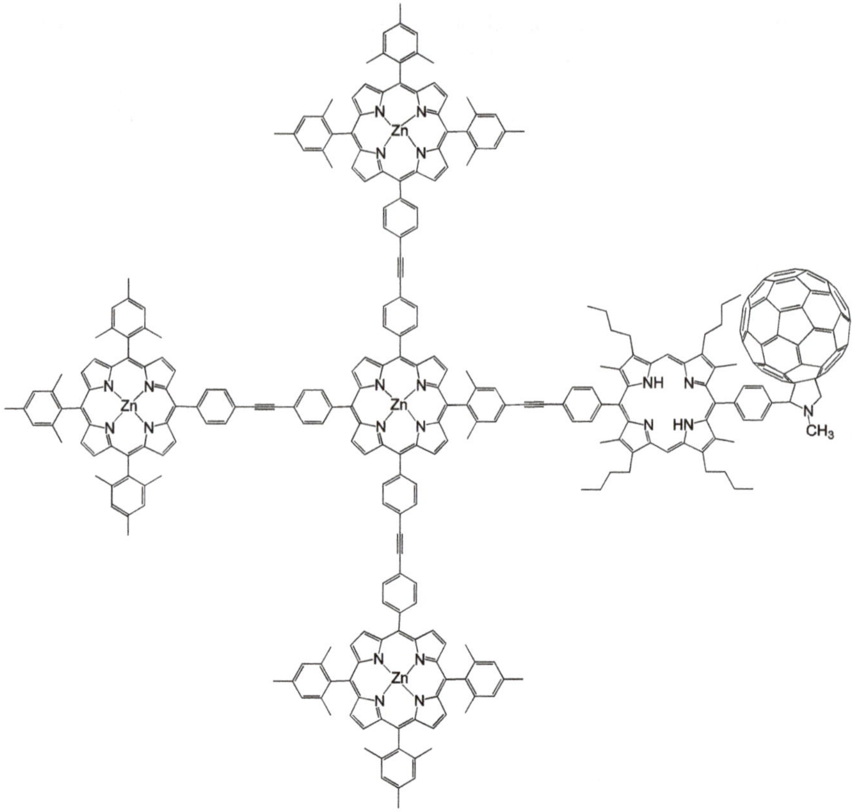

Fig. 8.2 Functional unit of an artificial antenna array linked to an artificial reaction center developed by Kuciauskas et al. (1999)

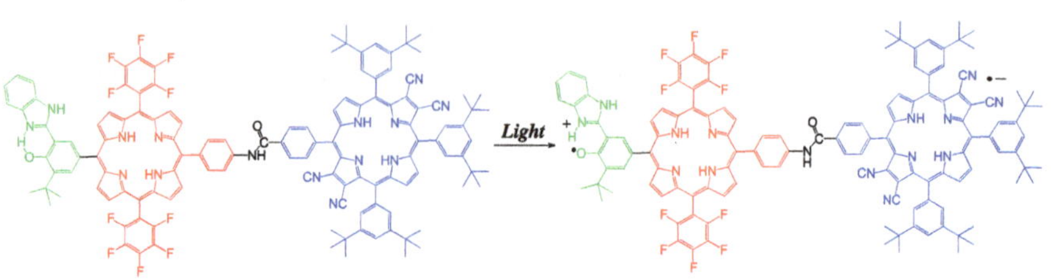

Fig. 8.3 Molecular structure of a triad which is composed of three covalently linked subunits mimicking the redox processes in Photosystem II. Upon illumination, sequential electron transfer (ET) and proton-coupled electron transfer (PCET) reactions yield the final charge-separated state BiH$^+$–PhO$^\bullet$–PF10–TCNP$^{\bullet-}$. See text for further explanations. According to Megiatto et al. (2012)

to simulate the specific interactions between P680, pheophytin and the TyrZ-His190 pair. The energetics of this triad was calculated in a way that the final charge-separated state is thermodynamically feasible of oxidizing water.

The tetracyanoporphyrin TCNP unit (blue) is a strong electron acceptor due to the electron withdrawing effects of the cyano groups at the β-positions of the tetrapyrrolic core. Two pentafluorophenyl groups ensure that the primary electron donor porphyrin (PF$_{10}$, red) is kept in the redox balance to transfer one electron to the TCNP acceptor after illumination. Oxidation of the benzimidazole-phenol (Bi-PhOH) secondary electron-donor component (green) is thermodynamically feasible by the formed PF10$^{\bullet+}$ cation. The Bi-PhOH has an intramolecular hydrogen bond between the phenolic proton and the lone pair of the nitrogen atom in the benzimidazole moiety. While Bi-PhOH is oxidized by PF10$^{\bullet+}$, the phenol is capable of transferring its proton to the benzimidazole group by a PCET mechanism. The neutral phenoxyl radical (E = 1.06 V vs. SCE) is mimicking the role of the TyrZ-His190 couple by being energetically able to oxidize water (E^0 = +0.58 V vs. SCE, pH 7).

A different attempt to couple light excitation to electron transfer mimicking the basic principles of PSII was carried out by Åkermark and co-workers (e.g., [13]), who synthesized simple model systems where a [Ru(bpy)$_3$]$^{2+}$-type photosensitizer (bpy = 2,2'-bipyridine) was covalently linked to a redox active moiety containing manganese or tyrosine. When [Ru(bpy)$_3$]$^{2+}$ is illuminated, an electron is transferred from the excited state of ruthenium(II) to an external acceptor, forming ruthenium (III). The generated hole is subsequently refilled by an intramolecular ET from the incorporated redox-active moiety. The choice of [Ru(bpy)$_3$]$^{2+}$ was based on its long-lived excited state lifetime, high chemical and thermal stability and easy functionalization procedures [14]. Flash photolysis experiments with the complex in Fig. 8.4 and electron paramagnetic resonance (EPR) spectroscopy confirmed that intramolecular ET occurs to yield a manganese (III) species. The process displayed showed a rate constant of $1.8 \cdot 10^5$ s^{-1}, following first-order kinetics. This was the first work demonstrating that coupled ruthenium-manganese systems are able to

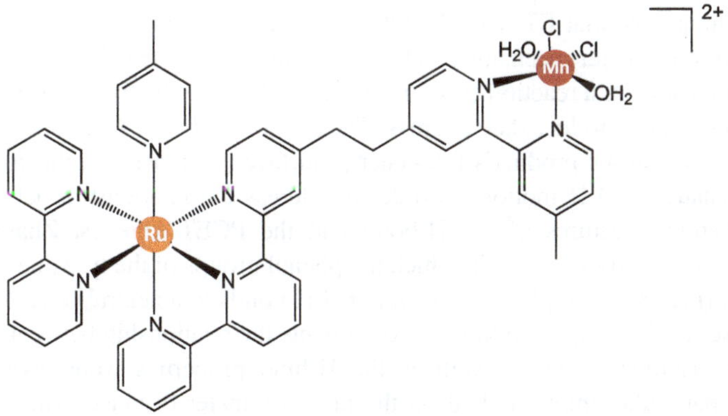

Fig. 8.4 Molecular structure of a mononuclear manganese complex covalently linked to a ruthenium photosensitizer for photoinduced electron transfer studies. According to Kärkäs et al. 2013

undergo intramolecular ET to the photooxidized ruthenium (III) center while an external electron acceptor is present.

8.1.2 Proton-Coupled Electron Transfer Studies

The remarkable quantum efficiency of Photosystem II has encouraged research groups also in the development of synthetic models to mimic its PCET reactions. These models initially focused on photoinduced ET in covalently coupled donor-bridge-acceptor (D – Br – A) systems (e.g., [15, 16]). Several models have also been developed over the years to simulate the tyrosine radical formation coupled to a Mn cluster, similarly to the water-splitting complex in PSII as described above (e.g., [17, 18, 19]) or the PCET reaction between P680$^{•+}$ and the Tyr$_Z$-His190, e.g. [20]. The assumption is made that TyrZ oxidation by P680$^{•+}$ occurs along the transfer of the phenolic proton to a hydrogen-bonded histidine residue, His190 [20]. Debates on the nature of this proton-coupled electron transfer (see Sect. 4.2.1) has inspired researchers to design model systems in which the tyrosine oxidation reaction can be studied more easily. Moore et al. (2008) reported on the synthesis of a bioinspired hybrid system, composing of a colloidal TiO$_2$ nanoparticles surface modified with BiP-PF$_{10}$, a photochemically active mimic of the photosynthetic chlorophyll-Tyr-His complex (see Fig. 8.3). Sjödin et al. 2000 designed a system consisting of a ruthenium photosensitizer linked to a tyrosine moiety (complex 1, Fig. 8.5) [21]. In the presence of an external acceptor, the Ru(II) complex can be photooxidized to Ru(III) on the nanosecond time-scale. In aqueous solutions, subsequent intramolecular electron transfer from the tyrosine residue follows on the microsecond time-scale, regenerating the Ru(II) sensitizer. Simultaneously, a tyrosine radical appears, which can be studied in optical and EPR experiments. The pK$_a$ of the tyrosine shifts from 10 to −2 upon oxidation, which coupled the process to a deprotonation reaction. Due its strong pH dependence, the authors demonstrate that ET from TyrOH to Ru(bpy)$_3$ has to occur as a concerted electron-proton transfer mechanism. This means that for the reaction to take place, the system must be in a reaction coordinate region allowing the free-energy surfaces for the reactant and product state to cross. The phenolic tyrosine OH bond is broken in this reaction and the product's free- energy surface is repulsive in the coordinate which translates to O-H motion. In order to address the relationship between the physico-chemical features of the H-bond and the PCET kinetics, Zhang et al. (2011) designed study systems, in which the phenol groups of the tyrosine linked to RuII(bpy)$_3$ (Fig. 8.5, complex 2) have internal H-bonds to a neutral base as in PSII [22]. In laser flash and quench kinetic experiments the relationship between the rate and kinetic isotope effects as well as the H-bind properties were investigated, whereas a particular emphasis laid on the proton transfer distance. This complex with a conjugated base attached to the phenol moiety demonstrated a higher rate of concerted PCET than in samples with corresponding non-conjugated ones, illustrating the significant effect of small geometric differences on the PCET rate and the

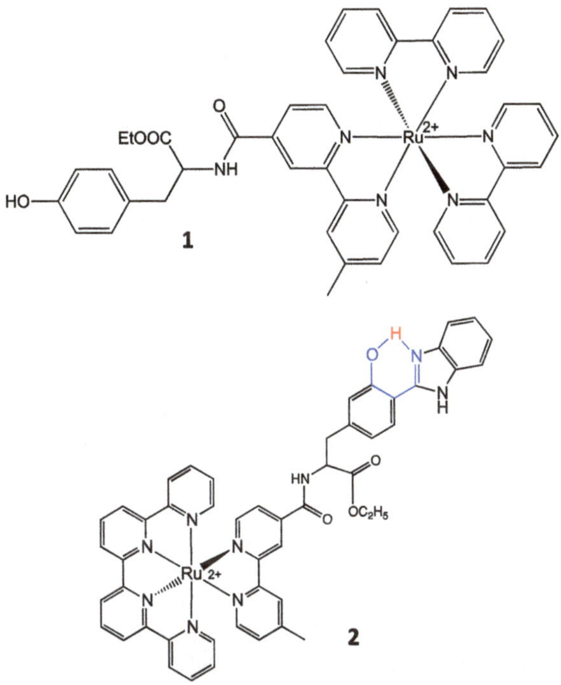

Fig. 8.5 Biomimetic systems mimicking the electron transfer from tyrosine to a photosensitizer. **1** Ru-tris-bipyridine absorbs light at 450 nm and when excited it transfers an electron to an external electron acceptor (metyl viologen). Tyrosine donates an electron to the photooxidized Ru(III) and is deprotonated. According to Sjödin et al. 2000. **2** Hydrogen-bonded tyrosine linked to a Ru(II) photosensitizer. The phenolic proton is H-bonded to an internal base (benzimidazyl group, blue). According to Zhang et al. 2011

importance of the proton transfer component and H-bonding in enzymes and their artificial counterparts.

8.1.3 Synthetic Mimics of the Mn$_4$CaO$_5$ Cluster

A key process for the production of solar fuels is the catalytic oxidation of water to molecular oxygen as accomplished by the Mn$_4$CaO$_5$ cluster in Photosystem II. It is therefore not surprising that researchers have tried to synthesise analogue manganese complexes for decades in order to elucidate the natural water-splitting mechanism, but also to design artificial mimics for realizing artificial photosynthesis. Pioneering work, where Mn oxides were successfully used as bio-inspired catalysts, was reported in the former USSR. As early as 1968, Glikman and Shchegoleva demonstrated oxygen evolution from aqueous suspensions of MnO$_2$ in the presence of the strong oxidant Ce^{4+} [23]. In 1981, the group of Shilov continued to work on this system and realized its model character for water oxidation in biological photosynthesis [24]. Long before the details of the OEC in PSII were known, Shilov and coworkers rightly assumed that manganese-centred redox chemistry in a Mn^{IV}-rich environment might be a very suitable process for water oxidation catalysts mimicking OER in PSII. They attributed the assumption to its ability to accumulate the four oxidation equivalents at the high redox potentials required for the reaction. Far away from any link to enzymatic water-oxidation chemistry were the studies carried out by electrochemists in Japan and Italy in the 1980s: the groups demonstrated that manganese oxides acted as electrocatalysts for the OER either as coatings on conductive surfaces or in the form of massive MnO_x discs/rods [25, 26]. These electrolysis reactions using MnO_x as anode materials were investigated at extreme pH conditions of acidic or alkaline electrolysers (pH ~ 0 or pH ~ 14, respectively) whereas the manganese oxides provided a sufficient stability. The application of MnOx as an anode material, enabled the possibility to reach water-oxidation current densities comparable to those observed for the oxides of Ru, Ir, Co and Ni.

Recent work from Kurz et al. (2016) with manganese(III) oxides has yielded in higher activity manganese materials than previously available. Solution-phase methods resulted in their preparations in crystalline and high surface area materials which operated as catalysts for the OER. Of particular interest of this work is the high activity obtained from manganese (III) oxides containing calcium; their high surface area preparations of the naturally occurring mineral Marokite ($CaMn_2O_4$) seemed to be more active for water oxidation than the calcium-free analogues [27].

With the recent atomic-resolution X-ray crystal structure of PSII, the focus of the Mn model chemistry has shifted to synthetic structures closely related to the water-splitting complex in PSII. Whereas it is difficult to synthetically imitate the exact geometry of nature's cuboidal Mn_4CaO_5 cluster, manganese complexes of various shapes have been synthesised and characterized [28, 29]. Only a very few of them, however, demonstrated water oxidation. The first example is shown in Fig. 8.6. The $[Mn_2^{III/IV}(terpy)_2(\mu - O)_2(OH_2)_2](NO_3)_3$ (terpy = 2,2';6',2''-terpyridine) catalyst consists of a Mn_2O_2 core which also resembles the Mn-di-μ-oxo units of the OEC [30]. The complex catalyses the OER in the presence of oxo-transfer oxidants such as ClO^- and HS_5^-. Mechanistic investigations revealed that the O_2 evolution rate saturates at oxidant concentrations above 13 000 equiv. with V_{max} = 2420 mol O_2 (mol complex)$^{-1}h^{-1}$ [29]. Beyond the protein scaffold in

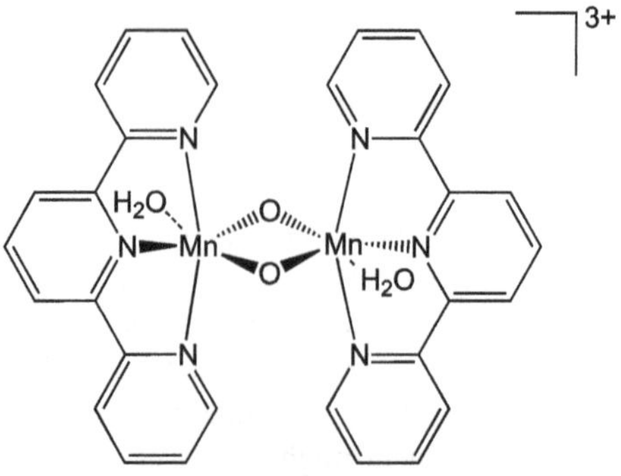

Fig. 8.6 Manganese-terpyridine dimer capable of oxidizing water. According to Limburg et al. 1999

PSII, the properties of the complex demonstrate the necessity of a tetranuclear system for tuning the redox potentials and maintaining the stability of the cluster.

This excursion into biomimetic model chemistry illustrates that it is possible to employ the tools of chemistry, biochemistry and materials science and the fundamental principles of photosynthetic energy conversion to design and build artificial photosynthetic systems which are not only model systems to reveal the secrets of natural photosynthesis, but which can use sunlight to generate fuels. The photosynthetic blueprint works due to an astoundingly fine-tuned molecular machinery—the artificial photosynthetic systems, however, require improvements in efficiency and durability before they can be considered for practical applications. Nevertheless, nature shows us that the challenge can be met.

8.2 The Photoelectrochemical Device: Utilizing Solar Energy for Water Oxidation, Hydrogen Production and CO_2 Fixation

Many approaches were undertaken so far to mimic the photosynthetic mechanisms of solar energy conversion. The final goal, however, is their integration into a fully functional solar-water splitting device which is operated as an 'artificial leaf' [31]. Due to the progress and existing research efforts in the development of a photoelectrochemical water-splitting device, this chapter focuses on its basic principles and current design trends.

Figure 8.7 summarizes PEC devices which have been reported so far, whereas the hydrogen evolution reaction can be exchanged with a CO_2 reduction reaction.

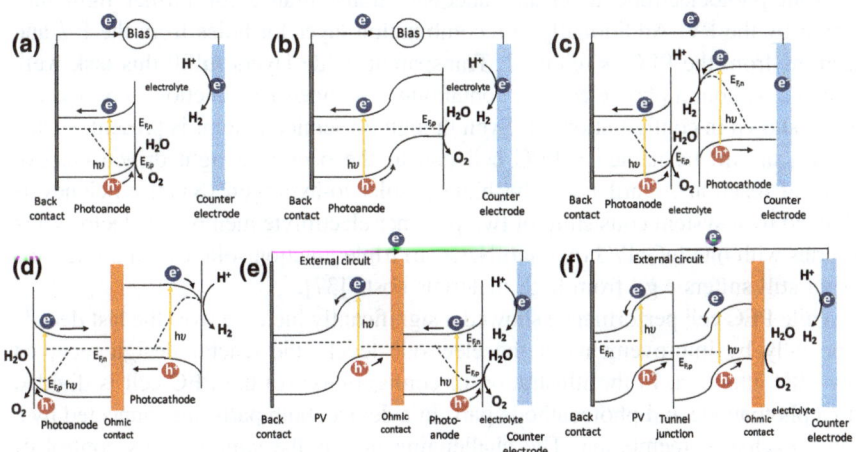

Fig. 8.7 PEC water-splitting device configurations: **a** single band gap light absorber; **b** heterojunction photoelectrode; **c** wired PEC tandem cell; **d** wireless PEC tandem cell; **e** PV-PEC tandem cell; **f** PV-electrolyser cell. According to Jiang et al. 2017

A basic photoelectrochemical water splitting device can be constructed using a single n-type or p-type semiconductor as a single band gap device (a) or from two, in series connected semiconductors, the so-called "dual band gap device", p/n-PEC (d). A dual band gap photoelectrolysis cell is of great advantage for obtaining efficient water splitting with currently available materials due to the ability to explore various combinations of smaller band gap semiconductor materials which have complementary absorption and stability characteristics (see Sect. 3.2, [32]). Configuration (b) illustrates a heterojunction photoelectrode PEC device. In these systems, the deposition of a secondary semiconductor onto a primary light absorber improves light absorption and charge separation [33]. Another approach is the utilization of in series-connected photocathodes and -anodes in a tandem configuration. This PEC cell contains two light absorbers: a photoanode for water oxidation and a photocathode for water reduction or CO_2 reduction, respectively. A conductive wire is used for the wired configuration (c). A transparent conductive substrate can also be employed as an electron-hole recombination layer for the wireless, tandem configuration in (d). In both cases, the conduction band minimum of the photoanode has to lie more negative than the valence band maximum of the photocathode to ensure complementary light absorption. In this arrangement, photons which are transmitted through the first material are absorbed by the second. A critical issue with this tandem configuration is the generation of high photovoltage under illumination. Furthermore, both sides have to maintain a similar current density when no external bias is applied.

Another cell type based on unbiased water splitting is achieved by coupling a PV cell with a semiconductor photoelectrode as shown in (e). In these systems, an external bias is applied to the photoelectrode by the PV material, composing of single or multiple junctions based on silicon [34], group III-V materials [35] or hybrid perovskites [36]. Preferably, wide band gap semiconductors are employed as the front photoelectrode to ensure adequate transmittance for further light harvesting by the PV. Additionally, a recombination layer for holes from the PV and electrons from the PEC is required. Transparent oxide layers fulfill this task well. A device without a photoelectrode containing a shielded PV junction (a buried PV cell) is shown in configuration (f). Even though no semiconductor is present, such a device can still be named a PEC cell due to the occurring light driven electrochemical reactions. Until now, the highest solar-to-hydrogen (STH) efficiency is achieved by a system consisting of two polymer electrolyte membrane electrolysers in series with one InGaP/GaAs/GaInNAs(Sb) triple junction solar cell, although the design still suffers e.g., from high materials costs [37].

While PEC cell performance showed a significantly increase over the last decade -especially by improvements at the electrode level—the reactor design received only little attention. In the illustration of configuration (c) the PEC cell is divided into a photoanode and photocathode part. In a device, both parts are connected by a proton exchange membrane. The challenging task is the simultaneous control of several processes and parameters: (i) electronic transport, (ii) ionic transport (i.e. proton transport), (iii) molecular transport and barrier function as well as (iv) optical transparency (photon management). In terms of electronic transport, efficient electrical conductivity between anode and cathode is essential. This can be achieved either externally by using an external wire or internally by direct electrode contact (e.g., configuration (d)). Furthermore, the condition of the electronic connection between the semiconductor and the catalytic sites responsible for oxidation and reduction reactions is important, imposing high demands on quality and continuity of electrode materials [38]. The ionic transport management comes into play in a PEC cell due to the proton transport from the anode to the cathode, comparable to the proton transport through channels in Photosystem II. Proton conduction is often more critical than electron conduction, since high ionic conductivity is more difficult to achieve: media which conduct ions have two to four orders of magnitude lower conductivity than the components conducting electrons in the PEC cell [38]. The achievement of efficient proton transfer can be realized either in liquid where the cell operation at high or low pH is an obvious way to increase conductivity (whereas changing the pH of the medium also influences the energy levels of the redox couples, semiconductors and catalytic centers) or in a solid electrolyte, possessing a high ionic conductivity. The proton conductivity of Nafion®, a commonly used sulfonated tetrafluoroethylene based copolymer can reach up to 10 S m^{-1} [39]. In the same way as in standard solid electrolytes, PEC cells would also benefit from materials with even higher conductivities while maintaining sufficient molecular barrier function. Furthermore, the proton transport distance is of importance: wired PEC cells with facing electrodes reach a higher efficiency due to shorter proton transfer distance. Molecular transport has been of interest particularly with respect to the low solubility of CO_2 and its availability at the catalytic sites of the photoelectrodes (see Sect. 6.2). Pressurized and cooled PEC cells have been considered [40, 41] as well as working under supercriticial conditions [42]. One way to overcome this obstacle is the operation of a PEC cell under gas or supercritical conditions, which provides sufficient amount of CO_2, but water might be limiting and proton conductivity is less evident. Furthermore, the isolation of the reaction product might be problematic. Furthermore, the production of oxygen at the photoanode will lead to the formation of gas bubbles, which increases the overpotential necessary to drive the reaction. The same accounts for the photo-cathode, where H_2 and/or CO2 reduction products are generated. To avoid the interference with gas bubbles, it is generally considered to operate the cell in vapor phase instead of liquid. Furthermore, PEC cells require the input of light, limiting the choice of materials. As an alternative to the arrangement of photoelectrodes in configuration (d), the electrodes can be placed side by side, although it likely reduces the overall efficiency since the illuminated surface area needs to be larger [38]. One consideration is the illumination of the PEC cell from both sites using mirrors, which on the other hand also complicates its design. As described in Sect. 4.4.1, nanostructured electrodes have proven to be an efficient strategy to reduce specular reflection and increase internal scattering for an optimal absorption of the incident light. Furthermore, the PEC cell casing should be transparent or contain a window for solar illumination. High-grade optical windows such as fused silica or quartz with anti-reflective coatings were already developed for PV panels, although one further difficulty with PEC cells is the chemical stability upon contact with the electrolyte due to the harsh reaction conditions.

A relatively new approach to gain understanding of the complex processes occurring inside a PEC cell and optimizing its design is taken by theoretical modelling. Simulation is an invaluable aid, providing directions for research efforts, assistance in the design, optimization and scale-up of PEC cells as well as the identification of bottle necks. For instance, Haussener et al. developed an adequate model based on mass and charge conservation, including Nernst-Planck description of species transport and Butler-Volmer equations for describing the electrochemical kinetics to simulate and compare different PEC cell geometries [39].

Research in the development of a biomimetic artificial leaf implemented by the design of a fully functional and efficient photoelectrochemical device for solar fuel production is a very active and popular research field. The PEC cell design is, however, a complex matter (as is the natural leaf) and involves simultaneous managing of a multitude of processes related to illumination, charge separation, electrical conduction, molecular transport, catalytic chemistry, substrate supply and reaction product recovery. A multidisciplinary effort is required to meet this challenge which also includes the necessity of a better and deeper understanding of natural photosynthesis by developing synthetic model analogues and by studying the natural system using latest developments and techniques in spectroscopy.

References

1. D. Gust, T.A. Moore, A.L. Moore, Mimicking photosynthetic solar energy transduction. Acc. Chem. Res. **34**, 40–48 (2001)
2. D. Gust, T.A. Moore, A.L. Moore, Solar fuels via artificial photosynthesis. Acc. Chem. Res. **42**(12), 1890–1898 (2009)
3. T.J. Meyer, Chemical approaches to artificial photosynthesis. Acc. Chem. Res. **22**, 163–170 (1989)
4. S. Fukuzumi, H. Imahori, Biomimetic electron-transfer chemistry of porphyrins and metalloporphyrins. Electron Transf. Chem. **2**, 927–975 (2001)
5. P.A. Liddell, D. Kuciauskas, J.P. Sumida, B. Nash, D. Nguyen, A.L. Moore, T.A. Moore, D. Gust, Photoinduced charge separation and charge recombination to a triplet state in a carotene-porphyrin-fullerene triad. J. Am. Chem. Soc. **119**, 1400–1405 (1997)
6. D. Gust, P. Mathis, A.L. Moore, P.A. Liddell, G.A. Nemeth, W.R. Lehman, T.A. Moore, R. V. Bensasson, E.J. Land, C. Chachaty, Energy transfer and charge separation in carotenoporphyrins. Photochem. Photobiol. **37S**, S46 (1983)
7. D. Gust, Very small arrays. Nature **386**, 21–22 (1997)
8. R.W. Wagner, J.S. Lindsey, A molecular photonic wire. J. Am. Chem. Soc. **116**, 11181–11193 (1994)
9. D. Kuciauskas, P.A. Liddell, S. Lin, T.E. Johnson, S.J. Weghorn, J.S. Lindsey, A.L. Moore, T.A. Moore, D. Gust, An artificial photosynthetic antenna-reaction center complex. J. Am. Chem. Soc. **121**, 8604–8614 (1999)
10. R. Berera, C. Herrero, L.H.M. van Stokkum, M. Vengris, G. Kodis, R.E. Palacios, H. van Amerongen, R. van Grondelle, D. Gust, T.A. Moore, A.L. Moore, J.T.M.A. Kennis, Simple artificial light-harvesting dyad as a model for excess energy dissipation in oxygenic photosynthesis. Proc. Natl. Acad. Sci. U.S.A. **103**, 5343–5348 (2006)

11. S.D. Straight, G. Kodis, Y. Terazono, M. Hambourger, T.A. Moore, A.L. Moore, D. Gust, Self-regulation of photoinduced electron transfer by a molecular nonlinear transducer. Nat. Nanotechnol. **3**, 280–283 (2008)

12. Megiatto J.D. Jr, A. Antoniuk-Pablant, B.D. Sherman, G. Kodis, M. Gervaldo, T.A. Moore, A.L. Moore, D. Gust, Mimicking the electron transfer chain in Photosystem II with a molecular triad thermodynamically capable of water oxidation. Proc. Natl. Acad. Sci. U.S.A. **109**(39), 15578–15583 (2012)

13. L. Sun, H. Berglund, R. Davydov, T. Norrby, L. Hammarström, P. Korall, A. Börje, C. Philouze, K. Berg, A. Tran, M. Andersson, G. Stenhagen, J. Martensson, M. Almgren, S. Styring, B. Åkermark, Binuclear ruthenium-manganese complexes as simple artificial models for Photosystem II in green plants. J. Am. Chem. Soc. **119**, 6996–7004 (1997)

14. M.D. Kärkäs, E.V. Johnston, O. Verho, B. Åkermark, Artificial photosynthesis: from nanosecond electron transfer to catalytic water oxidation. Acc. Chem. Res. **47**(1), 100–111 (2013)

15. M. Wasielewski, Photoinduced electron transfer in supramolecular systems for artificial photosynthesis. Chem. Rev. **92**(3), 435–461 (1992)

16. Y.K. Kang, I.V. Rubtsov, P.M. Iovine, J. Chen, M.J. Therien, Distance dependence of electron transfer in rigid, cofacially compressed, p-stacked porphyrin-bridge-quinone systems. J. Am. Chem. Soc. **124**, 8275–8279 (2002)

17. A. Magnuson, H. Berglund, P. Korall, L. Hammarström, B. Åkermark, S. Styring, L. Sun, Mimicking electron transfer reactions in Photosystem II: synthesis and photochemical characterization of a ruthenium(II) tris(bipyridyl) complex with a covalently linked tyrosine. J. Am. Chem. Soc. **119**(44), 10720–10725 (1997)

18. L. Sun, M.K. Raymond, A. Magnuson, D. LeGourrierec, M. Tamm, M. Abrahamsson, P.H. Kenéz, J. Mårtensson, G. Stenhagen, L. Hammarström, S. Styring, B. Åkermark, Towards an artificial model for Photosystem II: a manganese (II, II) dimer covalently linked to ruthenium (II) tris-bipyridine via a tyrosine derivative. J. Inorg. Biochem. **78**, 15–22 (2000)

19. T. Irebo, M.-T. Zhang, T.F. Markle, A.M. Scott, L. Hammarström, Spanning four mechanistic regions of intramolecular proton-coupled electron transfer in a $Ru(bpy)_{33}^{2+}$–tyrosine complex. J. Am. Chem. Soc. **134**(39), 16247–16254 (2012)

20. G.F. Moore, M. Hambourger, M. Gervaldo, O.G. Poluektov, T. Rajh, D. Gust, T.A. Moore, A.L. Moore, A bioinspired construct that mimics the proton coupled electron transfer between P680•+ and the TyrZ-His190 pair of Photosystem II. J. Am. Chem. Soc. **130**(32), 10466–10467 (2008)

21. M. Sjödin, S. Styring, B. Åkermark, L. Sun, L. Hammarström, Proton-coupled electron transfer from tyrosine-ruthenium-tris-bipyridine complex: comparison with tyrosine Z oxidation in Photosystem II. J. Am. Chem. Soc. **122**, 3932–3936 (2000)

22. M.-T. Zhang, T. Irebo, O. Johansson, L. Hammaström, Proton-coupled electron transfer from tyrosine: a strong rate dependence on intramolecular proton transfer distance. J. Am. Chem. Soc. **133**, 13224–13227 (2011)

23. T.S. Glikman, I.S. Shchegoleva, The catalytic oxidation of water by quadrivalent cerium ions. Kinet. Katal., 461–462

24. V.Y. Shafirovich, N.K. Khannanov, V.V. Strelets, Chemical and light-induced catalytic water oxidation. Nouv. J. Chim. **4**, 81–84 (1980)

25. M. Morita, C. Iwakura, H. Tamura, Anodic characteristics of massive manganese oxid electrode. Electrochim. Acta **24**, 357–362 (1979)

26. S. Trasatti, Electrocatalysis by oxides-attempt at a unifying approach. J. Electroanal. Chem. **111**, 125–131 (1980)

27. P. Kurz, Biomimetic water-oxidation catalysts: manganese oxides. Top. Curr. Chem. **371**, 49–72 (2016)

28. E.Y. Tsui, J.S. Kanady, T. Agapie, Synthetic cluster models of biological and heterogeneous manganese catalysts for O_2 evolution. Inorg. Chem. **52**, 13833–13848 (2013)

29. K.J. Young, B.J. Brennan, R. Tagore, G.W. Brudvig, Photosynthetic water oxidation: insights from manganese model chemistry. Acc. Chem. Res. **48**, 567–574 (2015)

30. J. Limburg, J.S. Vrettos, L.M. Liable-Sands, A.L. Rheingold, R.H. Crabtree, G.W. Brudvig, A functional model for O-O bond formation by the O_2-evolving complex in Photosystem II. Science **283**, 1524–1527 (1999)
31. D. Nocera, The artificial leaf. Acc. Chem. Res. **45**(5), 767–776 (2012)
32. M.G. Walter, E.L. Warren, J.R. McKone, S.W. Boettcher, Q. Mi, E.A. Santori, N.S. Lewis, Solar water splitting cells. Chem. Rev. **110**, 6446–6473 (2010)
33. C. Jiang, S. Moniz, A. Wang, T. Zhang, J. Tang, Photoelectrochemical devices for water splitting-materials and challenges. Chem. Soc. Rev. **46**, 4645–4660 (2017)
34. F.F. Abdi, L. Han, A.H.M. Smets, M. Zeman, B. Dam, R. van de Kroel, Efficient water-splitting by enhanced charge separation in a bismuth vanadate-silicon tandem photoelectrode. Nat. Comm. **4**, 2195–2202 (2013)
35. O. Khaselev, J.A. Turner, A monolithic photovoltaic-photoelectrochemical device for hydrogen production via water splitting. Science **280**, 425–427 (1998)
36. J. Luo, J.-H. Im, T. Mayer, M. Schreier, M.K. Nazeeruddin, N.-G. Park, S.D. Tilley, H.J. Fan, M. Grätzel, Water photolysis at 12.3% efficiency via perovskite photovoltaics and Earth-abundant catalysts. Science **345**, 1593–1596 (2014)
37. J.Y. Jia, L.C. Seitz, J.D. Benck, Y.J. Huo, Y.S. Chen, J.W.D. Ng, T. Bilir, J.S. Harris, T.F. Jaramillo, Solar water splitting by photovoltaic-electrolysis with a solar-to-hydrogen efficiency over 30%. Nat. Commun. **7**, 13237–13244 (2016)
38. J. Rongé, T. Bosserez, D. Martel, C. Nervi, L. Boarino, F. Taulelle, G. Decher, S. Bordiga, J. Martens, Monolythic cells for solar fuels. Chem. Soc. Rev. **43**, 7963–7981 (2014)
39. S. Haussener, C. Xiang, J.M. Spurgeon, S. Ardo, N.S. Lewis, A.Z. Weber, Modeling, simulation, and design criteria for photoelectrochemical water-splitting systems. Energy Environ. Sci. **5**, 9922–9935 (2011)
40. B. Kumar, M. Llorente, J. Froehlich, T. Dang, A. Sathrum, C.P. Kubiak, Photochemical and photoelectrochemical reduction of CO_2. Annu. Rev. Phys. Chem. **63**, 541–569 (2012)
41. E.V. Kondratenko, G. Mul, J. Baltrusaitis, G.O. Larrazábal, J. Pérez-Ramírez, Status and perspectives of CO_2 conversion into fuels and chemicals by catalytic, photocatalytic and electrocatalytic processes. Energy Environ. Sci. **6**, 3112–3135 (2013)
42. D.C. Grills, E. Fujita, New directions for the photocatalytic reduction of CO_2: supramolecular, $scCO_2$ or biphasic ionic liquid-$scCO_2$ systems. J. Phys. Chem. Lett. **1**, 2709–2718 (2010)

Chapter 9
Efficiency of Photosynthesis and Photoelectrochemical Cells

The 21st century confronts us with significant challenges, summarized by the UN Sustainability Goals. The forecast growth of our world population requires large increases in crop yields [1] and affordable, clean energy to meet the challenge of climate change. In the quest for an increase in crop production and within the context of biofuel production, questions concerning the efficiency of natural photosynthesis and its potential improvement arise. Similarly, the recent development of disparate technological approaches to convert solar energy into electricity and fuels such as photovoltaic (PV) cells, Photoelectrochemical Cells (PEC) cells and solar-thermal systems requires the adoption of a consistent approach to report energy-conversion efficiencies; a key metric that facilitates comparison of the performance of various approaches to solar energy conversion.

This chapter focusses on the definition and discussion of the energy-conversion efficiencies in natural photosynthesis and the ones obtained in solar water splitting cells. It outlines and discusses improvement strategies for both systems and finally addresses the question if the much-discussed solar conversion efficiencies in natural photosynthesis and photoelectrochemical cells are directly comparable.

9.1 The Photosynthetic Efficiency

In order to determine whether the photosynthetic efficiency can be improved, it is of primary interest to determine its theoretical maximum which could be reached under ideal conditions in C3 and C4 plants.

Oxygen-evolving photosynthesis uses only a limited part of the solar spectrum; photons below 400 nm and above 700 nm can only be used at low efficiencies—if at all. The photosynthetically active radiation (PAR) is therefore roughly 400–700 nm, representing 48.7% of the solar spectrum [2], translating to an average energy per mole photons of 205 kJ. Already at this point, 51.3% of the solar energy reaching the Earth's surface is not used (Fig. 9.1). Although a blue photon with a

© Springer International Publishing AG, part of Springer Nature 2018
K. Brinkert, *Energy Conversion in Natural and Artificial Photosynthesis*,
Springer Series in Chemical Physics 117,
https://doi.org/10.1007/978-3-319-77980-5_9

wavelength of 400 nm has 75% more energy than a red photon at 700 nm, the excited states of chlorophyll relax rapidly and photochemistry is carried out in the photosynthetic reaction centers with the energy of a red photon regardless of the originally absorbed wavelength. Assuming equal partitioning of photons between PSI and PSII during noncyclic electron transport, this results in 6.6% energy loss as heat, the 'photochemical inefficiency'. The charge separation process in the reaction centers dictates further on the thermodynamic limit: a charge separation event in Photosystem II requires about 176 kJ mol^{-1} [3] which is equal to the energy of a photon of its maximum absorption wavelength, 680 nm. For Photosystem I, the energy is about 171 kJ mol^{-1} ($\lambda = 700$ nm). On average, the energy loss between absorption and charge separation in the two photosystems is therefore about 31.5 kJ mol^{-1} ((205-(176+171)/2)kJ mol^{-1}). For this reason, only 63% of the energy of a red photon is used in the charge separation reaction and 37% is lost. Further energy losses are linked to electron and proton transport and the reduction of CO_2 to carbohydrate in the C3 cycle with additional losses in the C4-dicarboxylate cycle of C4 photosynthesis. C4 plants employ a more elaborated way of the more common C3 carbon fixation pathway since C4 overcomes the tendency of the enzyme RuBisCO to fix oxygen instead of carbon dioxide in the process of photorespiration. This is achieved by keeping the O_2 level in the working environment of RuBisCO works down. CO_2 is transferred via malate or aspartate from mesophyll cells to bundle-sheath cells. In these bundle-sheath cells, CO_2 release occurs by decarboxylation of the malate. C4 plants use a carboxylase for capturing more CO_2 in the mesophyll cells. In C3 photosynthesis, a minimum of 2 NADPH and 3 ATP is required to assimilate one molecule of CO_2 into carbohydrate and further on to generate 1 RuBP to complete the cycle. In the complete linear electron transport chain, 4 photons are required for the reduction of one molecule NADPH while a maximum of 6 protons is transferred into the thylakoid lumen (2 from water oxidation and 4 from plastoquinol oxidation by the cytochrome b_6f complex via the Q cycle). For the assimilation of one CO_2 into carbohydrates, two NADPH and 3 ATP are required, whereas 4 protons are necessary for the synthesis of 1 ATP. In total, 8 mol of red photons are needed to convert 1 mol of CO_2 into carbohydrate, translating to 874 kJ available energy for the conversion process. Therefore, the minimum energy expenditure in carbohydrate biosynthesis is 1-(477 kJ/874 kJ) or 10.78% (1C carbohydrate unit contains 477 kJ of energy). Prior to photorespiration and respiration Prior to photorespiration and respiration, the maximum energy conversion efficiency for a C3 plant is therefore given with 12.6% [3]. C4 plants such as maize and sugar cane require an additional 2ATP for the phosphorylation of pyruvate to phosphoenol pyruvate and therefore, a total number of 5 ATP to fix 1 CO_2 molecule. In C3 plants, oxygenation and photorespiratory metabolism represent a significant energy loss, essentially reducing the energy conversion efficiency to more than a half by 6.1%. Another source of energy loss is mitochondrial respiration. Ratios of respiratory CO_2 loss as a fraction of photosynthetic CO_2 uptake varies between 30 to 60%, depending on the crop type. The assumption is made in Fig. 9.1 that 30% accounts for the minimum energy loss due to respiration. The maximum solar energy conversion

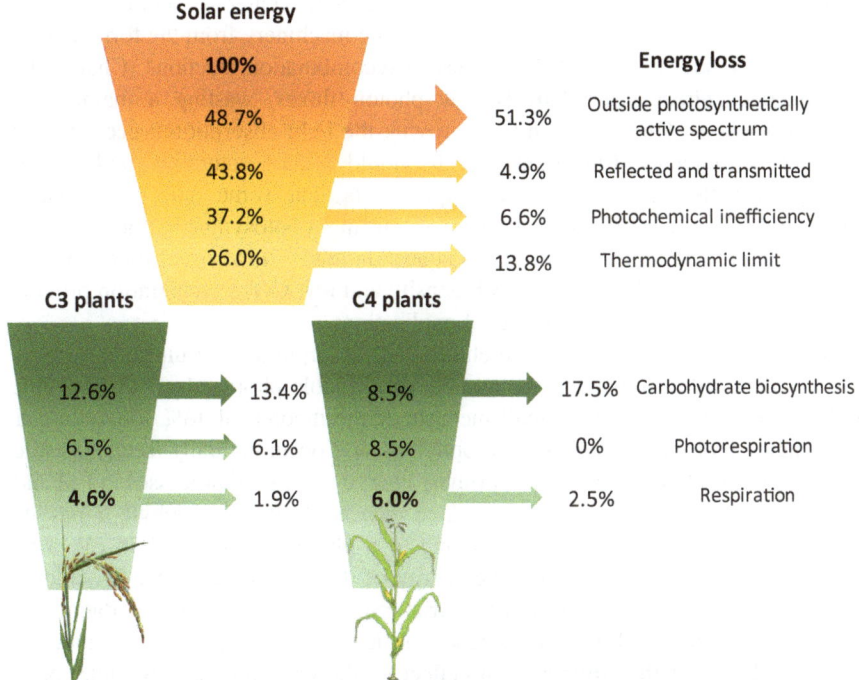

Fig. 9.1 Minimum energy losses in C3 and C4 plants showing the percentage remaining (inside the funnel) and percentage losses (right arrow) from an original 100% incident sunlight on a leaf to plant biomass. A leaf temperature of 30°C is assumed in an atmospheric [CO$_2$] of 387 ppm. According to Zhu et al. (2010)

efficiencies are therefore at 30°C 4.6% for C3 and 6% for C4 plants (also see Zhu et al. 2010 for more details on the calculations).

For photosynthetic microbes, the highest photosynthetic efficiency under controlled lab conditions is 3% for light-to-biomass [4], which is lower in industrial settings.

9.2   Optimizing Photosynthesis?

There are several approaches considered to 'engineer' and improve photosynthetic processes with respect to (i) light harvesting, (ii) photochemistry of the reaction center pigments, (iii) carbon fixation and (iv) the metabolic pathways [5].

The first step of photosynthesis represents light absorption by pigments in the antenna complexes which transfer exciton energy to the chlorophylls in the reaction centers. Due to the high number of antenna pigments, reaction center chlorophylls receive excitions over a significantly large surface area. Complex regulatory mechanisms, however, assure that when the reaction center is saturated, excess energy is converted to heat in order to protect the machinery from the formation of single oxygen species formed due to further recombination reactions (Chap. 7.1). This happens already at relatively low photon fluxes, wasting a majority of absorbed photons. One approach to improving the light absorption process would therefore be antennae truncation, which should reduce saturation and related shading obstacles. It is shown in experiments that some antennae mutants demonstrate indeed a slight improvement of the photosynthetic yields under specific, controlled conditions [6 7]. The question arises, however, if these specific conditions such as high light, high cell density and low CO_2 concentration provide real gains in energy terms for practical applications.

Furthermore, an extension of the chlorophyll absorption spectrum to 750 nm is considered in order to increase the number of available photons by 19% [8]. This could in principle result in a small increase of the theoretical 13% limit. Longer wavelength chlorophylls such as chlorophyll d and f occur naturally (see [9]) while chlorophyll d is already involved in primary charge separation processes in PSI and PSII in the cyanobacterium *Acaryochloris marina* [10]. Chlorophyll d has its absorption peak at 710 nm, corresponding to a photon energy of 1.75 eV. This energy is about 70 meV less than the energy of the photons absorbed by chlorophyll a (680 nm = 1.82 ev). Given the strict requirements for water oxidation, an increase in the probability of back-reactions leading to singlet oxygen induced damage is likely. Although there is no effect on the growth efficiency under optimized conditions [11], PSII becomes more susceptible to photodamage for variable light conditions. Far-red pigments, however, are used in the photosynthetic reaction centers in nature, although only in restricted environments [9]. Therefore, in order to apply this strategy, controlled growth conditions are required which (in case of biofuel production), lead again to higher energy costs [5].

As already discussed in Chap. 6.1, there are various ideas to improve the carbon fixation reaction, since the involved enzyme RuBisCO is slow and inefficient [12]. Clearly, there is scope for improvements, although most approaches still work within the 13% efficiency limit. Recently, an improved RuBisCO was reported to be expressed in *Synechocystis sp. PCC 6803*, showing a 2.9-fold activity increase accompanied by a 9% loss in CO_2/O_2 specificity [13]. Unfortunately, these cyanobacteria did not show any growth improvements and generated 25% less RuBisCO instead. This fact is consistent with the finding that changing the RuBisCO activity shows only little impact on the photosynthetic rate in *Synechocystis* [14], which is different under high CO_2 concentrations. Photorespiration is the pathway, which starts with RuBisCO oxygenation activity and recycles the product 2-phosphglycolate. This pathway represents a significant energy loss in C3 plants (see Fig. 9.1) and it has been shown that the reduction of photorespiratory losses can yield large growth improvements [15]. One approach is therefore to avoid RuBisCO and use alternative aerobic carbon fixation cycles such as the 3-hydroxypropionate bi-cycle [16]. These alternative cycles might have a lower requirement on ATP and could potentially improve the overall energy efficiency.

In terms of biofuel production, incorporating pathways into photosynthesising microbes is relatively straight forward since alkanes and e.g., isoprene and ethanol are made naturally [17–19]. The downside of the incorporation of pathways is the difficulty to specifically increase the product yield. Nevertheless, by co-expressing the mevalonic acid pathway, an isoprene producing strain was 2.5 times improved by Bentley et al. (2014) [20].

Presently, photosynthetic microbial biofuels are not viable in energy terms due to intrinsic inefficiencies in photosynthesis as discussed above (also see [5] and compare [21]), although evidence exists that increasing photosynthetic efficiency in crop plants can raise yield potential. Zhu et al. (2010) estimate that at least 50% improvement in the photosynthetic conversion efficiency will be needed to meet the doubled global productivity of grain crops which will likely be required over this century. To improve photosynthesis, however, a system approach is required, which is informed by coupled models able to correlate changes made in the chloroplast such as gene transfer and synthetic biology to yields in the field.

9.3   Efficiencies of Solar Water Splitting Cells

Fujishima's and Honda's report of light-driven photoelectrochemical water-splitting on single-crystal titanium in the absence of an applied bias ignited considerable interest in exploring solar water-splitting in practical means to generate fuels. Research efforts focused on finding other semiconductor materials which could yield in higher efficiencies. Much of the subsequent work included wide-band gap metal oxides and oxynitrides, possessing sufficient valence and conduction band levels for water oxidation. In 1975, Yoneyama et al. reported on the construction of a p-GaP/n-TiO$_2$ tandem combination which was able to generate H$_2$ and O$_2$ without the application of an external bias [22]. Consequently, in 1976, Nozik showed that this type of tandem structure consisting of a p- and n-type semiconductor could achieve much higher solar energy conversion efficiencies than a single photoelectrode [23]. The advances in single-junction and tandem-junction solar cells in the mid-1980s also reflected the increased efficiency of photoelectrochemical water-splitting cells. In the late 1990s, high-efficiency approaches based on III-V and Si/III-V monolithic tandem architectures were developed [24] which climaxed in the early 2000s with demonstrations of 12 and 18% solar-to-hydrogen efficiencies (STH, see following chapter for efficiency calculations) by Turner and co-workers [25] and Licht et al. (2000), respectively [26].

Since 2010, significant efforts have also been invested in replacing noble-metal electrocatalysts with those made from less expensive elements. The first demonstration by Nocera and co-workers reported a STH conversion efficiency of 2.5% for a completely integrated triple-junction a-Si cell employing hydrogen and oxygen evolution catalysts based on earth abundant elements [27].

Recently, Fountaine et al. (2016) calculated theoretical limiting efficiencies of water-splitting photoelectrochemical devices under ideal and realistic conditions and furthermore as well arbitrary intermediate conditions via parameter variation studies for single and dual junction photodiode systems [28]. The reported realistic limits consider the semiconductor absorption fraction, semiconductor external radiative efficiency (ERE), series resistance, shunt resistance and catalytic current exchange density and illustrate furthermore their correlation to light absorption, charge carrier transport and catalysis on the device performance.

The maximum single and dual junction efficiencies were calculated to be 30.6% at a band gap of 1.59 eV (threshold wavelength = 775 nm) and 40% with bandgaps of 0.52 and 1.40 eV, respectively (see Fountaine et al. 2016 for calculation details). Notably, the introduction of additional junctions did not result in any efficiency gains; the maximum triple junction efficiency for water-splitting is 28.3% which is significantly lower than the maximum dual junction efficiency. This result is in contrary to calculations of photovoltaic systems, which is explained by the authors that additional photovoltage beyond that required to kinetically split water does not result in an increase of efficiency. Moreover, the device photocurrent is reduced by an increased number of current-matched junctions which directly lower the efficiency (see (9.1)). The required output voltage of a water-splitting device is fixed at the water oxidation potential (1.23 eV) and therefore, the maximum conversion efficiency occurs for the semiconductor with a sufficient bandgap to generate 1.23 eV photovoltage. This translates to an effective bandgap of 1.59 eV due to the concentration and kinetic overpotentials needed to drive the reaction. For the theoretical simulations, however, an ideally constructed device with ideal photodiodes and catalyst performance is considered. Five parameters were further on identified which characterize a non-ideal photodiode and catalyst performance: (i) semiconductor absorption, (ii) semiconductor ERE, (iii) series resistance, (iv) shunt resistance and (v) catalytic exchange current densities. The combination of the latter one with the absorption fraction describes the efficiency of light absorption and catalysis. The carrier transport efficiency is described by three parameters, where ERE represents an intrinsic semiconductor quality, series and shunt resistance terms account for non-ideal charge transport. Series resistance spans the range from solution resistance to electrolyte transport, interfacial resistance at the semiconductor|catalyst interface and resistance to majority carrier low in the semiconductor, whereas shunt resistance results from partial shorting of diode junctions due to the ease with which liquid electrolyte can intercalate into pinholes of the semiconductor. Hereby, the maximum obtained single junction device efficiency was calculated to be 15.1% with a bandgap of 2.05 eV and 28.3% for a dual junction device with bandgaps of 0.92 and 1.59 eV. Considering single junction devices, a variation of the catalytic exchange current density resulted in both, the 'ideal' and 'real' case, in the largest device efficiency modulation and semiconductor bandgap. When the catalytic exchange current density is increased, kinetic overpotentials for the catalytic reaction increase as well whereas the semiconductor bandgap is still sufficient to carry out the water-splitting reaction. This leads to a precipitous drop in efficiency due to reduced solar spectrum conversion. Furthermore, the authors showed that a higher series resistant also has a significant effect on the single junction device efficiency, since it shifts the maximum power point to lower voltages and thus reduces the photocurrent near the maximum power point. In contrary to single junction devices, all five parameters have a distinct impact on the overall dual junction device efficiency. For instances, a poor catalyst performance is alleviated due to the series addition of two photovoltages. An increase in photovoltage is needed also here to compensate for the increase in kinetic overpotential. In a dual junction device, however, the photovoltage has to split between two semiconductors, resulting in a smaller increase in required bandgaps and therefore, having less effect on the overall device efficiency.

Interestingly, relatively small differences are observed between the theoretical efficiencies and experimental values obtained for an (1) 'ideal' and (2) 'real', high performance system such as (1) the dual junction $Pt_{black}|Si|AlGaAs|RuO_2$ device with an efficiency of 18.3% [26] and (2) the integrated dual junction $Rh|GaInP|GaInAs|RuO_2$ device with 14% efficiency [29]. The main difference between the calculated, theoretical maximum efficiency of a dual junction device of 40% (ideal conditions) and 28.3% (real conditions) and the experimentally obtained value of 18.3 and 14%, respectively, was attributed by the authors to a non-ideal bandgap selection in the initial design (1.1 and 1.6 eV for Licht et al. 2010; 1.26 and 1.78 eV for May et al. 2015). They calculated, however, the 'ideal' and 'real' maximum efficiencies of the two devices with 27.2% (24.5%) and 22.8% (20.5%), respectively, which—considering non-deal bandgap situation—are close to the values obtained theoretically.

9.3.1 Calculation of Solar-to-Chemical Conversion Efficiencies

In the context of the growing field of photoelectrochemical energy conversion and solar fuels, a number of metrics and definitions—often contradictory and not properly standardized—have been adopted for evaluating the performance of electrodes and systems [30]. It is therefore inevitable to establish a unifying terminology, mathematical expressions and experimental procedures to unify the comparison of photoelectrodes and solar fuel systems for evaluation with respect to other solar energy conversion technologies.

For a system generating products in the form of chemical fuels and/or electrical power the total output power P_O is defined as the sum of the output power contained in the chemical fuel, $P_{F,O}$. In the case that incipient output currents, I, are equal due to fuel and electricity production (i.e., the circuit elements are electrically connected in series), the relationship can be expressed as:

$$P_O = P_{f,o} + P_{e,o} = I \cdot (E_{f,o} + V_{e,o}) \tag{9.1}$$

Here, $E_{f,O}$ is the potential difference corresponding to the Gibbs free-energy difference between the two half-reactions of the fuels being produced. $V_{e,O}$ is defined as the output voltage of the electrical power portion of the total system output [30]. By definition, the efficiency of any process converting one type of energy into another one is the ratio of output power to input power. The general expression for the system efficiency, η, is therefore given by:

$$\eta = \frac{P_{f,o} + P_{e,o}}{P_S + P_{e,i}} \tag{9.2}$$

The total system input power, P_i, can be expressed as the sum of the electrical power input $P_{e,i}$ and/or the power from solar illumination, P_s:

$$P_i = P_S + P_{e,i} \tag{9.3}$$

For systems, such as PV or regenerative PEC cells which only produce electricity, the maximum output power P_{max} is expressed by the maximum operating current, I_{mp}, and voltage, V_{mp}. Thus, at maximum power, the efficiency of a photovoltaic or regenerative PEC system is given by:

$$\eta_{PV} = \frac{I_{mp} \cdot V_{mp}}{P_S} \tag{9.4}$$

In order to compare solar-fuels generators to a solar-electricity generating system, the Gibbs free energy of the fuel is considered according to Cordian et al. (2015) as the standardized energy content measure of the fuel. For a system producing only fuel as the output and using solar energy as the input, efficiencies can be calculated by setting the electrical power input and output terms to zero:

$$\eta_{STF} = \frac{P_{f,o}}{P_S} = \frac{A[cm^2] \cdot J_{op}[A\,cm^{-2}] \cdot E_{f,o}[V] \cdot \varepsilon_{elec}}{P_S[W]} \tag{9.5}$$

Here, J_{OP} is the operating current density, A the geometric area of the device and ε_{elec} is the Faradaic efficiency of the fuel production. For a solar-driven water-splitting system, the solar-to-hydrogen efficiency (STH), η_{STH}, is calculated using the difference in redox potentials of the hydrogen and oxygen evolution half-reactions ($E_{f,O} = 1.23$ V) to describe the Gibbs free-energy content of the products $H_{2(g)}$ and $O_{2(g)}$ which are formed under standard temperature and pressure conditions:

$$\eta_{STH} = \frac{A[cm^2] \cdot J_{OP}[A\,cm^{-2}] \cdot 1.23[V] \cdot \varepsilon_{elec}}{P_S[W]} \tag{9.6}$$

The most important parameter to characterise a PEC device is the solar-to-hydrogen efficiency (STH); this value can be used to rank all PEC devices against each other for benchmarking [31]. It describes the overall efficiency of a PEC water splitting device exposed to solar irradiance at an air mass number of AM

1.5 G (1.5 atmosphere thickness, corresponding to a solar zenith angle of z = 48.2°) under zero applied bias between working electrode (WE) and counter electrode (CE). This means that all energy necessary to drive the reactions originates from sunlight.

The application of a potential between working and counter electrode requires a new efficiency value separate from STH since the obtained value does not reflect a true solar-to-hydrogen conversion process (see [31] for more details). This commonly used procedure requires the definition of an "applied bias photon-to current efficiency" (ABPE) due to the fact that the application of a bias generally increases the current drawn from the device. ABPE measurement, however, can serve as a diagnostic tool in the development of new materials:

$$\text{ABPE} = \frac{j_{ph}[\text{mAcm}^{-2}] \cdot (1.23[\text{V}] - |V_b|) \cdot \varepsilon_{elec}}{P_S[\text{W}]} \tag{9.7}$$

Many values are generally reported for a PEC cell, whereas the STH efficiency is the most important one and great care must be taken to ensure that the measurement is carried out correctly with an accurate understanding of the effects of photoelectrochemistry. It is the only efficiency which can be used to determine the efficiency of water splitting H_2 production as a benchmark value.

9.4   Comparison of Photosynthetic and Photoelectrochemical Cell Efficiency?

A recurring topic in the performance discussion of photoelectrochemical water-splitting cells is their efficiency in comparison to natural photosynthesis which has also found attention in more recently published literature [32]. To evaluate the performance of PEC cells for solar water-splitting, a direct comparison to the solar energy conversion efficiency of plants makes little sense. The theoretical efficiencies of 4.6 and 6% for solar energy conversion in C3 and C4 plants, respectively, are very optimistic in comparison to observed experimental efficiencies, which are with a few exceptions about one-third of the theoretical value [2]. One reason for the lower field efficiency is frequent changes in the light intensity: at low light, more than 80% of the absorbed photosynthetically active photons can be used, whereas at 50% of full sunlight intensity (~ 1000 μmol m^{-2} s^{-1}), as little as 25% of the absorbed photons are used. At full sunlight intensity, this value falls to <10% [2] and protection mechanisms prevent photodamage (see Chap. 7.1). Solar-to-hydrogen efficiencies are usually reported with respect to the full AM 1.5 G solar spectrum, corresponding to an integrated light intensity of 100 mW cm^{-2}. This intensity corresponds to the solar irradiance measured after passing through the earth atmosphere and hitting the earth's surface at a solar zenith angle of 48.19°s. In efficiency discussions of photosynthesis, the full AM 1.5 G spectrum as reported by the American Society for Testing and Materials (ASTM, 280 nm to $\sim$2500 nm) is not considered. Therefore, a comparison of the photosynthetic efficiency to PEC cell efficiencies has to not only be based on the same solar spectrum, but it also has to take into account the different sunlight intensities at which both systems operate at their maximum conversion efficiency. This makes a direct comparison rather difficult. It is also notable that nature can adjust to light intensity variations by changing e.g., the antenna size system as previously discussed.

The energy conversion process in photosynthesis covers everything that a plant gets up to, day and night, during an annual cycle and the whole complex process, however, not to mention the plant's way of life, is certainly not a target for chemical mimicry. Although, even if the overall process might not look very efficient at a first instance, its quantum efficiency i.e., the percentage of absorbed photons giving rise to stable photoproducts, is almost 100%. Moreover, its water-splitting catalyst, the Mn_4CaO_5 cluster, is unbeaten with respect to its high turnover frequency number of up to 25–90 O_2 molecules released per second when compared to other catalysts consisting of earth abundant elements. Moreover, the low-energy pathway which extracts the electrons from water in a way that the enzyme can operate at a minimum electrical overpotential is difficult to mimic.

One further aspect is in this respect that we do not need energy in form of starch or sucrose; our energy demands are very different. We need transportable energy carriers with a high energy storage density such as hydrogen. Furthermore, we have a much higher demand energy: recent life-cycle assessments estimating the net energy implications of a hypothetical large-scale PEC hydrogen production facility were based on a total annual average production value of 1 GW hydrogen with a solar-to-hydrogen conversion efficiency device of 10% [33]. Already much lower STH efficiencies significantly affect the energy metrics.

Additionally, photosynthesis comprises various protection and repairing mechanisms; in fact, the D1 protein is the most rapidly turned-over protein in the thylakoid membrane [34]. Its degradation strongly depends on the incident light intensity and it can have a half-life of only 30 min [35]. The fact that photosynthesis can repair certain parts of its machinery or intrinsically synthesise new parts, increases its life time. Sathre et al. (2014) considered the aspect of life-cycle primary energy balance in the analysis of the hypothetical large-scale PEC hydrogen water oxidation facility in order to evaluate how much usable energy the facility provides during its life span, which was projected to be 40 years [33]. This very optimistic value is certainly outcompeted by nature: live oak trees can live for more than 500 years; some bristlecone pines are even thought to be more than 5000 years old.

Natural photosynthesis is a very well-tuned clockwork which operates perfectly for photosynthetic organisms and is well-conserved and protected in nature for more than 2 billion years. The above aspects also put the term energy conversion efficiency in a relative perspective: although, nature has a relatively low overall solar water oxidation efficiency, photosynthesis still operates more stable and reliable than any artificial biomimetic system today—it operates all over the planet and under various different conditions. Therefore, we can certainly learn about nature's efficiency compromises to protect its machinery and allow quick adjustments to environmental changes, but its actual solar-to-biomass conversion efficiency is not directly relevant for the design of efficient artificial photosynthesis systems.

References

1. L.H. Ziska, J.A. Bunce, H. Shimono, D.R. Gealy, J.T. Baker, P.C.D. Newton, M.P. Reynolds, K.S.V. Jagadish, C. Zhu, M. Howden, L.T. Wilson, Food security and climate change: on the potential to adapt global crop production by active selection to rising atmospheric carbon dioxide. Proc. Biol. Sci. **279**(1745), 4097–4105 (2012)
2. X.-G. Zhu, S.P. Long, D.R. Ort, Improving photosynthetic efficiency for greater yield. Annu. Rev. Plant Biol. **61**, 235–261 (2010)
3. X.-G. Zhu, S.P. Long, D.R. Ort, What is the maximum efficiency with which photosynthesis can convert solar energy into biomass? Curr. Opin. Biotechnol. **19**(2), 153–159 (2008)
4. A. Melis, Solar energy conversion efficiencies in photosynthesis: minimizing the chlorophyll antennae to maximize efficiency. Plant Sci. **177**, 272–280 (2009)
5. C.A.R Cotton, J.S. Douglass, S. De Causmaeker, K. Brinkert, T. Cardona, A. Fantuzzi, A.W. Rutherford, J.W. Murray, Photosynthetic constraints on fuels from microbes. Front. Bioeng. Biotechnol. **3**(36) (2015)
6. J.H. Mussgnug, S. Thomas-Hall, J. Rupprecht, A. Foo, V. Klassen, A. McDowall et al., Engineering photosynthetic light capture: impacts on improved solar energy to biomass conversion. Plant Biotechnol. J. **5**, 802–814 (2007)
7. H. Kirst, C. Formighieri, A. Melis, Maximizing photosynthetic efficiency and culture productivity in cyanobacteria upon minimizing the phycobilisome light-harvesting antenna size. Biochim. Biophys. Acta **1837**, 1653–1664 (2014)
8. M. Chen, R.E. Blankenship, Expanding the solar spectrum used by photosynthesis. Trends Plant Sci. **16**, 427–431 (2011)
9. M. Chen, M. Schliep, R.D. Willows, Z.-L. Cai, B.A. Neilan, H. Scheer, A red-shifted chlorophyll. Science **329**, 1318–1319 (2010)
10. T. Renger, E. Schlodder, The primary electron donor of Photosystem II of the cyanobacterium Acaryochloris marina is a chlorophyll d and the water oxidation is driven by a chlorophyll a/ chlorophyll d heterodimer. J. Phys. Chem. B **112**, 7351–7354 (2008)
11. S.P. Mielke, N.Y. Kiang, R.E. Blankenship, M.R. Gunner, D. Mauzerall, Efficiency of photosynthesis in a Chl d-utilizing cyanobacterium is comparable to or higher than that in Chl a-utilizing oxygenic species. Biochim. Biophys. Acta **1807**, 1231–1236 (2011)
12. Y. Savir, E. Noor, R. Milo, T. Tlusty, Cross- species analysis traces adaptation of Rubisco toward optimality in a low-dimensional landscape. Proc. Natl. Acad. Sci. U.S.A. **107**, 3475–3480 (2010)
13. P. Durão, H. Aigner, P. Nagy, O. Mueller-Cajar, F.U. Hartl, M. Hayer-Hartl, Opposing effects of folding and assembly chaperones on evolvability of RuBisCO. Nat. Chem. Biol. **11**, 148–155 (2015)
14. Y. Marcus, H. Altman-Gueta, Y. Wolff, M. Gurevitz, Rubisco mutagenesis provides new insight into limitations on photosynthesis and growth in *Synechocystis PCC6803*. J. Exp. Bot. **62**, 4173–4182 (2011)
15. R. Kebeish, M. Niessen, K. Thiruveedhi, R. Bari, H.-J. Hirsch, R. Rosenkranz et al., Chloroplastic photorespiratory bypass increases photosynthesis and biomass production in *Arabidopsis thaliana*. Nat. Biotechnol. **25**, 593–599 (2007)

16. J. Zarzycki, V. Brecht, M. Müller, G. Fuchs, Identifying the missing steps of the autotrophic 3-hydroxypropionate CO_2 fixation cycle in *Chloroflexus aurantiacus*. Proc. Natl. Acad. Sci. U.S.A. **106**, 21317–21322 (2009)
17. J. Han, E.D. McCarthy, W.V. Hoeven, M. Calvin, W.H. Bradley, Organic geochemical studies, II. A preliminary report on the distribution of aliphatic hydrocarbons in algae, in bacteria, and in a recent lake sediment. Proc. Natl. Acad. Sci. U.S.A. **59**, 26–33 (1968)
18. P. Lindberg, S. Park, A. Melis, Engineering a platform for photosynthetic isoprene production in cyanobacteria, using *Synechocystis* as the model organism. Metab. Eng. **12**, 70–79 (2010)
19. J. Dexter, P. Fu, Metabolic engineering of cyanobacteria for ethanol production. Energy Environ. Sci. **2**, 857–864 (2009)
20. F.K. Bentley, A. Zurbriggen, A. Melis, Heterologous expression of the mevalonic acid pathway in cyanobacteria enhances endogenous carbon partitioning to isoprene. Mol. Plant **7**, 71–86 (2014)
21. H. Michel, The nonsense of biofuels. Angew. Chem. Int. Ed. **51**, 2516–2518 (2012)
22. H. Yoneyama, H. Sakamoto, H. Tamura, A photoelectrochemical cell with production of hydrogen and oxygen by cell reaction. Electrochim. Acta **20**, 341–345 (1975)
23. A.J. Nozik, p-n photoelectrolysis cells. Appl. Phys. Lett. **29**(3), 150–153 (1976)
24. J.W. Ager, M.R. Shaner, K.A. Walczak, I.D. Sharp, S. Ardo, Experimental demonstrations of spontaneous solar-driven photoelectrochemical water splitting. Energy Environ. Sci. **8**, 2811–2824 (2015)
25. O. Khaselev, J.A. Turner, A monolithic photovoltaic-photoelectrochemical device for hydrogen production via water splitting. Science **280**, 425–427 (1998)
26. S. Licht et al., Efficient solar water splitting, exemplified by RuO_2-catalysed AlGaAs/Si photoelectrolysis. J. Phys. Chem. B **104**, 8920–8924 (2000)
27. S.Y. Reece, D.G. Nocera, Proton-coupled electron transfer in biology: results from synergistic studies in natural and model systems. Annu. Rev. Biochem. **78**, 673–699 (2009)
28. K.T. Fountaine, H.-J. Lewerenz, H.A. Atwater, Efficiency limits for photoelectrochemical water-splitting. Nat. Commun. **7**, 13706–13715 (2016)
29. M.M. May, H.-J. Lewerenz, D. Lackner, F. Dimroth, T. Hannappel, Efficient direct solar-to-hydrogen conversion by in situ interface transformation of a tandem structure. Nat. Comm. **6**, 8286–8272 (2015)
30. R.H. Coridan, A.C. Nielander, S.A. Francis, M.T. McDowell, V. Dix, R.H. Chatman, N. Lewis, Methods for comparing the performance of energy conversion systems for use in solar fuels and solar electricity generation. Energy Environ. Sci. **8**, 2886–2901 (2015)
31. Z. Chen, H.N. Dinh, E. Miller (eds.), *Photoelectrochemical Water Splitting* (Springer Briefs in Energy, New York, Heidelberg, Dordrecht, London, 2013)
32. R.E. Blankenship, *Molecular Mechanisms of Photosynthesis*, 2nd edn. (Oxford, Wiley-Blackwell, 2014)
33. R. Sathre et al., Life-cycle energy assessment of large-scale hydrogen production via photoelectrochemical water splitting. Energy Environ. Sci. **7**, 3264–3278 (2014)
34. M. Edelman, A.K. Mattoo, D1-protein dynamics in photosystem II: the lingering enigma. Photosynth. Res. **98**, 609–620 (2008)
35. E. Kanervo, P. Mäenpää, E.M. Aro, Localisation and processing of the precursor form of photosystem II protein D1 in *Synechocystis* 6803. J. Plant Physiol. **142**, 669–675 (1993)

Chapter 10
Conclusion: Towards the Realization of an Artificial Leaf

The dream of realising artificial photosynthesis can probably be best compared to the dream of flying at the turn of the 19th century; early aviation pioneers looked at birds, the function of the wings, the tail, fuselage and the aerodynamics of flying to come up with biomimetic aeroplane design features. Birds, however, also flap when flying, build nests and fly south for the winter—we do not care about these habits when crossing the Atlantic in an Airbus A380.

Mimicking natural photosynthesis translates to the same: we have to identity the specific principles and reactions which are potentially useful for an artificial system converting solar energy into for us useful energy rich products. Certainly, stability and earth abundance of the involved materials and compounds are aspects in the general discussion as well as economic feasibility and the operation of the device in regions with less solar irradiance, but various key points can be identified from the natural template. An efficient PEC device for solar water-splitting coupled to H_2 production and/or CO_2 reduction requires first of all light absorbers which generate a sufficient electrochemical potential upon light absorption to drive the catalytic reaction of water-splitting and hydrogen production/CO_2 reduction at a low overpotential with linked catalysts. Semiconductor-electrocatalyst systems as discussed here, are promising systems in this respect, although functional systems are mostly based on rare and costly elements such as Pt for H_2 production. Further research efforts are required to combine efficient, stable and earth abundant catalysts with semiconducting light absorbers. This is not an easy task, given the multielectron and multiproton processes involved. The high number of oxidation states in the Mn_4CaO_5 cluster might be something to consider in the design of artificial water-splitting catalysts, although Mn as a central element might not be the most obvious choice. Nature's Mn cluster is well-protected by several protein subunits at the heart of Photosystem II; several amino acid residues are able to stabilise the cluster in various stages of the catalytic cycle. We design isolated systems, in which the catalyst is in most cases also exposed to sunlight and the electrolyte and therefore, requires a much higher intrinsic stability. Promising artificial systems

© Springer International Publishing AG, part of Springer Nature 2018
K. Brinkert, *Energy Conversion in Natural and Artificial Photosynthesis*,
Springer Series in Chemical Physics 117,
https://doi.org/10.1007/978-3-319-77980-5_10

which have recently been developed are based on nickel, iron and cobalt which also posses a high number of oxidation states.

Photodamage due to high light intensities, photocorrosion and the production of singlet oxygen species are also obstacles in artificial systems which need to be avoided; one way forward is the application of protection layers such as TiO_2 which are already employed in semiconductor-electrocatalyst systems. Additional sensoric filters could be further on employed to prevent high light damage of the PEC cell.

Furthermore, substrate transfer to the catalyst and product transfer away from the catalytic center needs to operate effectively; the construction of channel systems as employed by nature in proximity of the Mn_4CaO_5 cluster might not be feasible from an engineering point of view, but their composition and operation principle might be useful in the design of new PEC cell membranes. Nature also employs an electron relay, Tyr_Z, between the photochemistry at P680 and the water oxidation site at the Mn_4CaO_5 cluster, participating directly in the catalytic activity of PSII. This relay undergoes proton-coupled electron transfer to a nearby histidine residue when transferring electrons from the catalytic center to oxidized $P_{680}^{\bullet+}$. It is an interesting feature which can also directly respond to pH changes around the cluster. Artificial catalysts and fully integrated catalyst-light absorber systems function optimally at a distinct pH value with little tolerance with respect to changes in electrolytic ion concentration and moreover, water purity. An electron relay is an interesting concept, but it might not be crucial for artificial systems since the system can be optimized to operate at a certain pH value and the device can be built in a way that changes in temperature or pressure do not affect the catalytic reaction and electrolyte concentration and pH maintain the same.

It is probably also not very wise to mimic nature's way of CO_2 fixation; again, this process was developed to serve the energy needs of a photosynthetic organism building its biomass based on starch and sucrose. If renewable fuels should be competitive to fossil fuels, our solar-driven CO_2 fixation process needs to be much more efficient and being able to produce specific products upon request. Intensive research has just started in this area during the last decade which demonstrates already great potential.

Although it seems that we can already draw many conclusions from studies of natural photosynthesis for the design of artificial, biomimetic systems, we have not found the ultimate solution for an efficient, artificial solar energy conversion system yet. Moreover, there are still many questions related to the natural process left, to the fore, the structure and function of the Mn_4CaO_5 cluster during catalysis. It becomes evident that interdisciplinary research has to continue hand in hand in both fields in order to further the development of a solar-driven device which converts water into hydrogen and oxygen while simultaneously reducing CO_2 to a valuable fuel. The complexity of natural photosynthesis is certainly not something we would like to mimic, but understanding its mechanisms and principles can guide us the way. Moreover, it teaches us that nature has almost unbelievable design skills which not only deserve our veneration, but which also need to be protected—for us and for our planet.

Index

A

ABPE. *See* applied bias photon-to-current efficiency
Adenosine triphosphate (ATP), 3, 6, 112, 114
ALD. *See* atomic layer deposition
Antennae. *See* light-harvesting antennae
Applied bias photon-to current efficiency, 119
Artificial photosynthesis, 1, 97
Atomic Layer Deposition (ALD), 93

B

Benson-Bassham-Calvin cycle. *See* Calvin cycle
Bio-inspired catalysts, 103
Butler-Volmer equation, 45, 47, 108

C

C3 plants, 111, 112, 114, 119
C4 plants, 111, 112, 119
Calvin cycle, 6, 7
Carbon fixation pathway, 112
Carotenoids, 15, 42, 89
Channel architecture, 56
Charge recombination, 28, 40–42, 89, 98
Charge separation, 1, 4, 12, 15, 17, 19, 27, 29, 35, 98, 106, 108, 112, 114
Chemical Vapour Deposition (CVD), 93
Chlorophyll a, 11, 17, 99, 114
Chloroplast, 3, 10, 39, 77, 115
Chromophore, 12, 14, 97
CO2 fixation, 6, 75, 77, 124
Corrosion, 1, 71, 92, 93
CVD. *See* chemical vapor deposition
Cytochrome b6f complex, 112

D

Dark reactions (photosynthetic), 4, 6

E

EET. *See* electronic excitation transfer
Electrocatalyst, 47, 50, 62, 63, 75, 82, 84, 123, 124
Electrolyser, 63
Electron donor-acceptor dyads, 98
Electronic excitation transfer, 12
Electron Paramagnetic Resonance (EPR), 39, 60, 101
Electron transfer, 3, 4, 11, 14, 15, 17, 19, 33, 35, 36, 40, 41, 43, 49, 56, 62, 63, 65, 66, 79, 80, 82, 88, 94, 97, 98, 102, 124
Energy conversion efficiency, 20, 82, 105, 111, 115
EPR. *See* electron paramagnetic resonance
ET. *See* electron transfer
Eukaryotic organisms, 10, 14
Exchange current density, 45, 46, 64, 116
Excitation energy transfer, 10, 11, 89, 98
External Radiative Efficiency (ERE), 116

F

Förster Resonance Energy Transfer (FRET), 12
Fermi level, 22, 23, 50, 62, 93
Ferredoxin-NADP reductase, 5
Franck-Condon principle, 43
FRET. *See* Forster resonance energy transfer

G

Gerischer, Heinz, 21, 23, 70

© Springer International Publishing AG, part of Springer Nature 2018
K. Brinkert, *Energy Conversion in Natural and Artificial Photosynthesis*,
Springer Series in Chemical Physics 117,
https://doi.org/10.1007/978-3-319-77980-5

H
Haber Bosch process, 55
Helmholtz layer, 23, 43
HER. *See* hydrogen evolution reaction
High valence (HV) scheme, 59
Hydrogen evolution reaction, 21, 55, 69, 94,
      105

I
Ideal diode equation, 47

K
Kinetic isotope effect, 102
Kok cycle, 18, 40, 59, 60

L
LHC. *See* light harvesting complex
Light-harvesting antennae, 10, 14
Light harvesting complex, 10, 14
Light reactions (photosynthetic), 3
Low valence (LV) scheme, 59

M
Marcus theory, 33, 36, 41, 49, 83
Marokite, 104
Minority-carrie, 26, 51
Mn4CaO5 complex. *See* oxygen evolving
      complex

N
NADPH, 3, 5, 6, 80, 112
Nafion®, 107
Nanostructured photoelectrodes, 50
N-type semiconductor, 22, 23, 26, 27, 46, 115

O
OEC. *See* oxygen-evolving complex
OER. *See* oxygen evolution reaction
Overpotential, 22, 28, 45, 47, 52, 56, 63–65,
      69, 79, 80, 82, 116, 117
Oxygen evolution reaction, 21, 27, 62, 63, 65,
      66, 68, 87
Oxygen-evolving complex, 5, 58, 69
Oxygenic photosynthetic organisms possess
      several, 89

P
P680, 4, 11, 15, 17, 18, 39, 40, 56, 88, 89, 100,
      124
P700, 12, 16, 39
PCET. *See* proton-coupled electron transfer
Perovskites, 63, 106
Photoanode, 21, 25, 27, 47, 55, 87, 92, 93, 106
Photocathode, 21, 22, 26, 47, 92, 106

Photocurrent, 48–50, 92, 94, 116
Photodamage, 10, 88, 89, 99, 114
Photoelectrocatalysis, 20, 21, 62, 71
Photoelectrochemical cell, 1, 21, 22, 49, 55,
      111, 119
Photoelectrochemical CO$_2$reduction. *See* CO2
      fixation
Photoelectrode, 21, 25, 29, 49, 50, 63, 80, 92,
      93, 106, 115
Photoexcitation, 17, 26, 27
Photoluminescence, 49
Photoprotection, 87, 99
Photosynthetically Active Radiation (PAR),
      12, 111
Photosynthetic model systems, 1, 97
Photosynthetic reaction center, 9, 10, 12, 20, 35
Photosystem I, 3, 4, 15, 39, 87, 112
Photosystem II, 3, 4, 17, 35, 36, 39, 41, 65, 87,
      89, 100, 103, 107, 123
Phylloquinones, 15, 16
Porphyrin, 80, 84, 97–99, 101
production, 120
Prokaryotic organisms, 10, 14
Protection layers, 92
Protection mechanisms, 1, 87, 119
Proton-coupled electron transfer, 36, 40, 66,
      82, 100, 102, 124
Proton exchange membrane, 106
PSI. *See* Photosystem I
PSII. *See* Photosystem II
P-type semiconductor, 22, 25, 26, 94, 106
Purple bacteria, 9, 13, 14, 42
PV junction, 106
Pyridine, 80, 81

Q
QA/QA-• couple, 41
Quantum efficiency, 48, 49, 102, 120

R
Radiation damage, 56, 60
Redfield theory, 13
Redox cofactors, 17, 33, 35
Ribulose-1,5-bisphosphate, 7, 76
Ribulose-1,5-bisphosphat-carboxylase. *See*
      rubisco
Rubisco, 7, 76–78, 82, 112
Rubp. *See* ribulose-1,5-bisphosphate
Ru(bpy)2+, 66

S
Schottky barrier contact, 23
Semiconductor/ liquid interface, 21
Single junction device, 117

Singlet oxygen, 42
Solar-to-hydrogen efficiency, 22, 26, 28, 110, 118
Solar-water splitting cell, 23
Space charge region, 23, 48
S states, 18, 60, 62
STH. *See* solar-to-hydrogen efficiency
Superoxide dismutase, 90
Surface passivation, 26
Surface states, 48, 49

T
Tandem configuration, 22, 106
Titanium dioxide (TiO2), 20, 27, 54, 92–94, 96, 102, 115, 124
TOF. *See* turnover frequency
Transition metal oxides, 27, 28, 55
Turnover frequency, 69, 82, 120
Two-dimensional electronic spectroscopy (2DES), 14

Tyr$_Z$. *See* tyrosine
TyrD. *See* tyrosine
Tyrosine, 39, 40

V
Volcano plot, 65, 70

W
Water oxidation, 18, 21, 22, 25, 27, 28, 53, 55, 56, 60–63, 65–69, 88, 93, 95, 99, 103–105, 109, 112, 114–116, 124
Water-splitting reaction. *See* water oxidation

X
X-Ray diffraction (XRD), 56
X-Ray Free Electron Laser (XFEL), 56

Z
Z-scheme, 4, 5